Adaptation, Learning, and Optimization

Volume 17

About this Series

The role of adaptation, learning and optimization are becoming increasingly essential and intertwined. The capability of a system to adapt either through modification of its physiological structure or via some revalidation process of internal mechanisms that directly dictate the response or behavior is crucial in many real world applications. Optimization lies at the heart of most machine learning approaches while learning and optimization are two primary means to effect adaptation in various forms. They usually involve computational processes incorporated within the system that trigger parametric updating and knowledge or model enhancement, giving rise to progressive improvement. This book series serves as a channel to consolidate work related to topics linked to adaptation, learning and optimization in systems and structures. Topics covered under this series include:

- complex adaptive systems including evolutionary computation, memetic computing, swarm intelligence, neural networks, fuzzy systems, tabu search, simulated annealing, etc.
- machine learning, data mining & mathematical programming
- hybridization of techniques that span across artificial intelligence and computational intelligence for synergistic alliance of strategies for problem-solving.
- aspects of adaptation in robotics
- agent-based computing
- autonomic/pervasive computing
- dynamic optimization/learning in noisy and uncertain environment
- systemic alliance of stochastic and conventional search techniques
- all aspects of adaptations in man-machine systems.

This book series bridges the dichotomy of modern and conventional mathematical and heuristic/meta-heuristics approaches to bring about effective adaptation, learning and optimization. It propels the maxim that the old and the new can come together and be combined synergistically to scale new heights in problem-solving. To reach such a level, numerous research issues will emerge and researchers will find the book series a convenient medium to track the progresses made.

More information about this series at http://www.springer.com/series/8335

Raja Humza Qadir

Self-Sufficiency of an Autonomous Reconfigurable Modular Robotic Organism

Springer

Raja Humza Qadir
dSPACE GmbH
Paderborn
Germany

ISSN 1867-4534 ISSN 1867-4542 (electronic)
ISBN 978-3-319-10288-7 ISBN 978-3-319-10289-4 (eBook)
DOI 10.1007/978-3-319-10289-4

Library of Congress Control Number: 2014948069

Springer Cham Heidelberg New York Dordrecht London

Printed on acid-free paper

Springer is part of Springer Science+Business Media (www.springer.com)

"To my beloved Parents"

Foreword

Today, robots are widespread and can be found in nearly any area: In our factories they weld, sort, and assemble. In our libraries they pick books, they autonomously store and deliver items in logistic centers. Robots explore the surface of distant planets and at home they vacuum clean our living rooms or mow our lawns. This has become possible due to a rapid progress in technology. However, the examples listed above are not that smart as it may seem at a first glance. All the mentioned robots have been specialized to perform a certain task which in most cases is tedious for human labor.

What is lacking to robots today is the ability to adapt to varying conditions. This is necessary in cases where human intervention is difficult to achieve or too dangerous. Search and rescue scenarios and the exploration of remote terrain are very often quoted in this context. Moreover, one tries to increase the autonomy of robots so that once given a task they perform without a human's assistance. Autonomy is absolutely required when robots are supposed to operate in a swarm, a large group of single entities that performs much superior compared to a single member of the group. Insect colonies are the archetype of this strategy that builds upon "swarm intelligence", a seemingly intelligent behavior evolving from a rather simple set of rules that swarm members follow.

Bringing this idea one step further, the swarm members can physically join to become a larger robot, i.e., create a robotic organism, just as our bodies consist of many body cells. This type of robots, on which scientists have yet not been working for long, is called "autonomously reconfigurable" robots. Some examples exist, which have evolved from a number of scientific projects.

The aim of the book that you hold in your hands right now is to increase the autonomy of such reconfigurable robots by increasing their energetic autonomy. Robots, just like any other machine, or biological creature, are dependent on the admission of energy to operate. Hence, by implementing smart energy distribution and dissipation strategies one can considerably increase the autonomy of a robotic organism.

Prof. Dr.-Ing. Oliver Scholz
University of Applied Sciences
Saarbrücken, Germany

Preface

In recent years, getting inspiration from simple but complex biological organisms, several advances have been seen in autonomous systems to mimic different behaviors that emerge from the interactions of a large group of simple individuals with each other and with the environment. Among several open issues a significantly important issue, not addressed so far, is the *self-sufficiency*, or in other words, the energetic autonomy of a modular robotic organism. This feature plays a pivotal role in maintaining a robotic organism's autonomy for a longer period of time.

To address the challenges of self-sufficiency, a novel dynamic power management system (PMS) with fault tolerant energy sharing is proposed, realized in the form of hardware and software, and tested. The innate fault tolerant feature of the proposed PMS ensures power sharing in an organism despite docked faulty robotic modules. Due to the unavailability of sufficient number of real robotic modules, a simulation framework called *Replicator Power Flow Simulator* is devised for the implementation of application software layer power management components. The simulation framework was especially devised because at the time of writing this work no simulation tool was available that could be used to perform power sharing and fault tolerance experiments at an organism level. The simulation experiments showed that the proposed application software layer dynamic power sharing policies in combination with the distributed fault tolerance feature in addition to self-sufficiency are expected to enhance the robustness and stability of a real modular robotic organism under varying conditions.

Acknowledgments

In the name of Allah, the most Gracious, the ever Merciful.

Time flies. The first day in Germany and days spent at University of Saarland, Saarbrücken and Fraunhofer Institute for biomedical engineering (IBMT), St. Ingbert, are still fresh in memory. Today, at the time of writing this book, I see many people around me whose continuous support I always regard and therefore using the opportunity I would like to explicitly gratitude them all.

First, I would like to thank my supervisor Prof. Dr. Oliver Scholz for giving me an opportunity to pursue doctoral studies at Microsystems Department, IBMT. His valuable guidance and criticism during the time has always helped me to develop new concepts and foster my research aptitude. He gave me ample freedom in decision- making and has always supported my initiatives in the project. I would like to thank Dr. Thomas Velten, Head of Department at IBMT, for his kindness during my stay at IBMT. I am also thankful to my colleagues in the group, especially Timo Koch and Nicolas Schumann, for providing technical support in developing and testing the electronics used for experiments in this book.

I shall always remember the great time spent with REPLICATOR and SYMBRION project fellows. In particular, I am grateful to Prof. John Timmis and Lachlan Murray, from University of York, United Kingdom, for giving me an opportunity to learn and develop new concepts in the area during my visit at Department of Electronics, University of York, UK.

And lastly, I am overwhelmed with gratitude to my father, my mother (may Allah rest her soul in peace), parents in-law, my wife and kids (Hafsah, Abeer and Eshal) for their continuous support during the whole journey.

Contents

Chapter 1
Introduction

In recent years, several advances have been seen in autonomous systems to mimic different behaviors that emerge from the interactions of a large group of simple individuals with each other and with the environment, termed *swarm behavior*. The term *swarm behavior* is assigned to the identical behavior of a large population of a specie that as a whole show complex pattern, flexibility and robustness. It was first used by Beni and Wang (1989), in the context of cellular robotic systems. In particular, it focuses on the collective behavior that evolves from the local interactions of the individuals with each other and with their environment (Dorigo 2009). Examples of such species are, ant and termite colonies, schools of fish, flocks of bird, herd of animals, etc. Some human artifacts also fall in the domain of swarm intelligence, notably multi-robot systems, and certain computer programs, written to tackle optimization and data analysis problems.

Getting inspiration from simple but complex biological organisms, like, slime mold, the robotic research community introduced a new breed of robotic system in the field of swarm robotics, composed of multiple homogeneous and heterogeneous "re-configurable robotic modules". A re-configurable robotic module can be defined as an artificial standalone entity having multiple tools, e.g., docking mechanism, camera, locomotive and dexterous drives, sensors, on-board power source, etc., to perform a variety of tasks individually and collectively that may include, self-sufficiency, self-healing, self-adaptability, self-organization, etc. Over the time, modular re-configurable robots that were only able to cooperate and collaborate logically in the swarm, now own the ability to extend their collaboration through physical means, i.e., by physically docking with each other. The term *re-configurable* is the characteristic feature of such systems to form or adapt different morphologies, which distinguishes them from a traditional robotic system.

Unlike a traditional robotic system, a re-configurable modular robotic organism combines the features and abilities of multiple robotic modules to achieve the objectives that are beyond the capabilities of a single robotic module. For instance, through mutual collaboration, physical and logical, multiple robotic modules may adapt a morphology to overcome different challenges in the environment, e.g., crossing an obstacle higher than the height of a robotic individual, energy harvesting from a power socket or recharge station mounted high in the wall, building a bridge, climbing

R.H. Qadir, *Self-Sufficiency of an Autonomous Reconfigurable Modular Robotic Organism*, Adaptation, Learning, and Optimization 17,
DOI 10.1007/978-3-319-10289-4_1

on a wall, etc. The physical collaboration between the robotic modules in an organism brings multiple degrees of freedom, but also involves multiple challenges or issues to address from their control and management perspective. The collective operations in a modular robotic organism are in fact dependent on the dynamic collaboration between the robotic modules in terms of dynamic resource sharing, e.g., computational power, sensing ability, load distribution, power sharing, etc. This in turn requires the adaptation of several system parameters not only to achieve self-sufficiency but also to sustain their aggregation in the organism. In the autonomy of a modular robotic organism a significantly important factor is its *self-sufficiency* or energetic autonomy. Up to now, the term self-sufficiency has only been used for single autonomous robots, but not for a modular robotic organism, composed of multiple individual robots. According to McFarland and Boesser (1993), it is the ability of a robotic system to autonomously find fuel and refuel itself from the environment. At an organism level, the complexity of the task multiplies with the number of robotic modules in an organism.

This research work in distributed autonomous systems particularly addresses an important system parameter that plays a critical role in maintaining the autonomy of a modular robotic organism for a longer period of time, namely, *self-sufficiency*.

1.1 Problem Description

Different principles in artificial autonomous systems have been devised to mimic the behavior of biological systems, observed in nature, for instance, the division of labor, trophallaxis, task sharing, leaving pheromone traces, etc. But, it is still challenging to establish both physical and virtual collaboration between multiple autonomous robotic module of a modular robotic organism. In the past, to keep the energetic autonomy of an autonomous robotic system in a swarm the problem of self-sufficiency has been explored and presented in a variety of scenarios, for instance, harvesting energy using biological means, e.g., sugar (Wilkinson 2000), digesting slugs (Ieropoulos et al. 2005b), and through artificial means, like fuel cells (Kelly et al. 2000), solar panels (Boletis et al. 2006; Landis and Jenkins 1997), recharging from stationary power stations (Munoz et al. 2002), swapping of battery packs (Ngo 2008), etc. But, so far the said principle has not been explored from the perspective of a modular robotic organism.

The self-sufficiency principle in a similar way for a modular robotic organism can be interpreted as the collective ability of autonomous robotic modules in an organism to find fuel and refuel. To accomplish this objective collectively, the robotic modules in a modular robotic organism are therefore must be remained operational till it successfully docks to an available recharge station in the arena. In this regard, to keep the physical collaboration between the robotic modules in a modular robotic organism despite their varying individual energetic status, the hardware design of the robotic modules must support power sharing and fault tolerance to withstand abrupt endogenous and exogenous faults and failures. Studying the platform designs of existing state-of-the-art reconfigurable robotic systems like SuperBot (Salemi et al. 2006) and ATRON (Lund et al. 2005) revealed that their solution although support power sharing but lack certain design features that are considered by the author as vital for dynamic power sharing between the robotic modules of an organism. For example, from the perspective of power sharing, a robotic module must be able to measure and control the current flowing from its battery pack to its on-board system components and to the docked robotic modules. Depending on a system's implementation, the current flow measurements at different components may allow a robotic module to dynamically adapt its behavior during different behavioral states in a modular robotic organism. In addition, the existing state-of-the-art reconfigurable robotic systems platform designs lack fault tolerance at the hardware layer that is essential to keep the physical collaboration of the robotic modules despite faults and failures in a modular robotic organism.

Considering the morphology of a modular robotic organism, in addition, the system design of an autonomous robotic module must bear features that allow it to control its power sharing behavior in conjunction with the docked robotic modules. In other words, the application software components of a robotic module must have an access to the innate system features, embedded in the hardware design, in order to cope different situations in a modular robotic organism, in particular, non-uniform or varying energetic status of the robotic modules and collective behavior of robotic modules in the presence of faults or component failures. So far, no such research work has been seen or conducted that explore the behavior of a modular robotic organism from the perspective of self-sufficiency.

1.2 Original Contributions

The detailed contributions of the work can be summarized as following:

- Firstly, to address the open issues concerning the self-sufficiency of a modular robotic organism, a behavioral modeling approach is used to define the self-sufficient behavior of a modular robotic organism. The devised behavioral model captures the behavioral states an organism may have to encounter to accomplish the self-sufficiency task in different scenarios. In this regard, to keep the physical collaboration of the docked robotic modules, the behavioral states highlighted the conditions in which an organism has to adapt different modes of power sharing between the robotic modules.
- To address the challenges of dynamic power sharing in different behavioral states an application software layer control system is proposed. In this regard, getting inspiration from the homeostasis phenomenon found in living beings, the term *artificial energy homeostasis* is defined in the context of a modular robotic organism.
- For the implementation of the concept, after a detailed analysis of the system requirements and limitations of a robot platform, in particular REPLICATOR robotic modules, a dynamic power management system with fault tolerant energy

sharing is proposed and developed. The proposed system for the sustenance of collaborative operations allows a robotic module to share its on-board energy reserve in the organism despite *faulty robotic modules*. In addition, it provides control over the inward and outward current flow in a robotic module, i.e., through its docking sides.

- To measure the devised system's efficiency and its behavior in different operating conditions with real robotic modules low level control software is then developed. The developed control software provides an access to control the system innate functionalities, power sharing and fault tolerance, at the application software layer. In addition, an external interface for system analyses is developed to record the current flow measurements through different system components of a robotic module.
- Due to the unavailability of sufficient number of real robotic modules, a simulation framework is devised to simulate and explore the effects of dynamic power sharing between multiple robotic modules in two exemplary organism morphologies. The simulation framework named *Replicator Power Flow Simulator*, consists of two parts: the front-end and the SPICE[1] simulation engine. The simulation front-end is custom designed in LabWindows/CVI. It provides a graphical user interface to visualize and configure the different parameters effecting the power sharing among the robotic modules of an organism. In addition, it is used to code the proposed application software layer power sharing strategies that are required to simulate the dynamic power flow in a modular robotic organism under different configurations. At the back-end, the integrated SPICE simulation software from Linear Technology—LTSpice IV—is used to obtain realistic power flow measurements in the organism.
- Following the artificial energy homeostasis definition, application software layer power management components required for dynamic power sharing in a modular robotic organism, are then defined. The proposed application software layer power management components of a robotic module include proactive power management, morphology graph, energy distribution graph and power sharing polices. The *Replicator Power Flow Simulator* is then used to explore the effects of initial energy distribution, the mode of locomotion and dynamic power sharing in two different organism morphologies. The obtained simulation results are then discussed from the perspective of self-sufficiency of a modular robotic organism.
- Lastly, from the perspective of fault tolerance, different kinds of faults are identified at an individual and a modular robotic organism level using a fault tree analysis approach. To develop a collective fault tolerant behavior at the application software layer, an exogenous and two endogenous fault conditions are simulated at different system components of the robotic modules in an organism. In this regard, for the detection and isolation of a faulty component or a robotic module in an organism, a fault detection and identification and a fault isolation algorithm is proposed at the application software layer of a robotic module. The fault tolerant behavior of the robotic modules in the organism, during power sharing, with the proposed algorithms is then simulated using the LTSpice simulation software.

[1] Simulation program with integrated circuit emphasis.

1.3 Organization of the Book

This book is organized in six chapters as: Chap. 2 provides a brief description of the background and the related work. In the start, few examples of swarm behaviors found in nature that served as the primary source of inspiration for swarm robotics are discussed. The chapter then briefly reviews the system design and features of couple of state-of-the-art re-configurable robotic platforms that support power sharing. It then introduces the design of a re-configurable robotic platform; namely REPLICATOR (Kernbach et al. 2008). The proposed power management system with fault tolerant energy sharing is designed, developed and integrated in the electronic design of REPLICATOR robotic modules as a part of this research work. Later, some of the bio-inspired techniques applied in swarm robotics to gain energy autonomy are reviewed. And lastly, it briefly describes the simulation tools used in the work.

Chapter 3 presents the core concept that is conceived to address the issues related to self-sufficiency of a modular robotic organism. In the beginning it presents the open issues at a modular robotic organism from the perspective of self-sufficiency. To model the self-sufficient behavior of a modular robotic organism, firstly, the energetic modes of a self-sufficient robotic module are defined using a finite state machine. On the basis of energetics of a robotic module, the behavior model of a self-sufficient robotic module in a robot swarm is then defined. By combining the individual behavior of self-sufficient robotic modules it then defines the self-sufficient behavior of a modular robotic organism. The chapter then presents the concept of dynamic power management in a modular robotic organism. In this regard, getting inspiration from the homeostasis phenomenon found in biological systems, it introduces a novel concept of artificial energy homeostasis to achieve self-sufficiency at a modular robotic organism level. The following publications was produced from the contents of this chapter.

- R. Humza, O. Scholz, M. Mokhtar, J. Timmis and A. Tyrrell. "Towards energy homeostasis in an autonomous self-reconfigurable modular robotic organism", In *Proceedings of the 2009 Computation World: The First International Conference on Adaptive and Self-Adaptive Systems and Applications*, pages 21–26, 2009.
- R. Humza and O. Scholz. Book Chapter: "Energy autonomy and energy harvesting in reconfigurable swarm robotics". In Symbiotic Multi-Robot Organisms: Reliability, Adaptability, Evolution, P. Levi and S. Kernbach, editors, pages 116–135. Springer-Verlag, 2010. ISBN 978-3-642-11691-9.
- R. Humza and O. Scholz. "A case study on self-sufficiency of individual robotic modules in an arena with limited energy resources". In *ADAPTIVE 2011, The Third International Conference on Adaptive and Self-Adaptive Systems and Applications*, pages 29–35, 2011.

Chapter 4 covers the implementation details of the proposed solution, both at the hardware and application software layer. In the beginning, the hardware design considerations from the perspective of a robotic module's platform design are described in detail. In this regard, the factors that influence the design of a dynamic power management system are explored in detail, that include it's electronic architecture, system power budget calculation, choice of system source voltage, and platform specific constraints. The chapter then presents the detail description of the proposed power management system with fault tolerant energy sharing and its design considerations. The proposed power management system has been developed and realized as a real microprocessor based hardware together with a dedicated firmware. From the perspective of application software layer implementation, firstly, it presents the devised simulation framework and its implementation details. And, then the details of the proposed application software layer power management components of a robotic module required for dynamic power sharing in a modular robotic organism, and the application software layer fault tolerance at an organism level are presented. At the end, a short summary concludes the presented work. The following publications include parts of the contents that are presented in this chapter.

- S. Kernbach, E. Meister, O. Scholz, R. Humza, J. Liedke, L. Ricotti, J. Jemai, J. Havlik and W. Liu. "Evolutionary robotics: The next-generation-platform for on-line and on-board artificial evolution" In *IEEE Congress on Evolutionary Computation, 2009. CEC'09*, pages 1079–1086, 2009.
- S. Kernbach, O. Scholz, K. Harada, S. Popesku, J. Liedke, R. Humza, W. Liu, F. Caparrelli, J. Jemai, J. Havlik, E. Meister and P. Levi. "Multi-Robot Organisms: State of the Art". In *ICRA10, workshop on "Modular Robots: State of the Art", Anchorage, 2010.*
- S. Kernbach, F. Schlachter, R. Humza, J. Liedke, S. Popesku, S. Russo, T. Ranzani, L. Manfredi, C. Stefanini, R. Matthias, Ch. Schwarzer, B. Girault, P. Alschbach, E. Meister and O. Scholz. "Heterogeneity for increasing performance and reliability of self-reconfigurable multi-robot organisms", In *workshop on Reconfigurable Modular Robotics, IROS-11*, San Francisco, 2011.

Chapter 5 is dedicated to the presentation of the experimental results obtained with the proposed hardware and the application software features in the simulation framework. In this regard, the experimental results are presented, obtained after integrating the proposed power management system in the hardware design of the real robotic modules, during artificial trophallaxis, power sharing and a fault condition. In the second half of the chapter, the application software, implemented in the simulation framework, is tested by simulations. The simulation experiments are broadly divided into two parts: power sharing and fault tolerance. The power sharing simulation experiments include the simulations of dynamic power sharing in two exemplary organism morphologies. The fault tolerance simulation experiments include the results obtained by adapting the procedure defined in the fault detection and identification and fault isolation algorithms, during different fault situations. At the end, a brief summary concludes the chapter.

Finally, Chap. 6 presents the conclusion with some ideas for future work.

Chapter 2
Background and Related Work

Abstract This chapter briefly reviews the background and the related work in swarm robotics from the perspective of self-sufficiency. It starts with swarm intelligence and provides few examples of swarm behaviors found in nature that serve as the source of inspiration for the advances in swarm robotics. In this regard, it presents the salient properties of robotic modules in a robot swarm and reviews the design of existing state-of-the-art re-configurable robotic platforms that support power sharing. It then introduces the design features of heterogenous REPLICATOR re-configurable robotic modules. The chapter then covers some of the bio-inspired techniques applied in swarm robotics to gain energy autonomy and briefly introduces the fault tolerance, in the context. And lastly, it briefly describes the simulation tools used in this research work.

2.1 Swarm Intelligence

The word "swarm" is usually designated to a population of apparently disorganized moving individuals that has the tendency to form clusters for different purposes without any centralized control. In nature, such a behavior, the purposeful collaboration of autonomous individuals in a group, can be seen in different species, e.g., ant colonies, bees, birds flocking, fish schooling, bacterial growth, etc. To study the collective behavior of apparently simple but complex individuals, Beni and Wang (1989) were the first who used the term "swarm intelligence" in their work of cellular robotic systems. *Swarm intelligence* focuses on the emergent collective behavior of independent individuals that results from the local interaction of group or swarm members among each other and with the environment without external supervision. Mark (1994), who developed swarm models for application in artificial life has formulated five basic properties of swarm intelligence:

- **Proximity principle**: the swarm members should be able to carry out simple computations concerning space and time,
- **Quality principle**: the swarm members should be able to evaluate the interactions between themselves and with the environment,

R.H. Qadir, *Self-Sufficiency of an Autonomous Reconfigurable Modular Robotic Organism*, Adaptation, Learning, and Optimization 17,
DOI 10.1007/978-3-319-10289-4_2

- **Diverse response principle**: the interactions among the swarm members must be diverse and distributed instead of concentrated on too narrow behavior,
- **Stability principle**: the swarm members in a swarm are generally less prone to repetitive fluctuations in the environment and therefore their behavior does not oscillate,
- **Adaptability principle**: the swarm is sensitive to rapid changes happening in the environment that may require a change in its behavior.

Bonabeau et al. (1999) defines the term swarm intelligence as,

> "*any attempt to design algorithms or distributed problem-solving devices inspired by the collective behavior of social insect colonies and other animal societies.*"

Later, Bonabeau and Meyer (2001) further explained the term swarm intelligence as,

> "*Social insects work without supervision. Their teamwork is largely self-organized, and coordination arises from the different interactions among the individuals of the colony. Although these interactions might be primitive, for instance, one ant merely following the trail left by another, taken together they lead to efficient solutions to difficult problems, such as, finding the shortest route to a food source among myriad possible paths. The collective behavior that emerges from a group of social insects has been dubbed swarm intelligence.*"

The swarm intelligence principles help to understand and model complex systems through simple rules that are then applied to solve problems in different areas. For instance, the "ant colony optimization" (ACO) that was initially proposed by Dorigo and Caro (1999), Dorigo et al. (1991, 1996). It is a meta-heuristic optimization algorithm that is applied to find approximate solutions to difficult combinatorial optimization problems that can be reduced to finding optimal paths through graphs. The ACO technique mimics the pheromone laying phenomenon found in natural ant colonies to build solutions by moving the artificial ants on the problem graph so the future artificial ants can build better solutions. This is done in an iterative process where the good solutions found by the ants of an iteration should guide the ants of the following iterations. Another related meta-heuristic method, which was introduced by Kennedy and Eberhart (1995), is known as "particle swarm optimization" (PSO). PSO is inspired by social behavior of birds in a flock and fish schooling and optimizes the problem by moving a population of possible or candidate solutions, also called as particles, in the search space. The movement of these particles is then directed to the best particles (evolved over time) found in the search space. It is applied in a variety of continuous optimization problems such as, training of artificial neural networks (ANN), finite element updating, etc. Other commonly known methods include, "stochastic diffusion search" (SDS) by Bishop (1989), and "river formation dynamics" (RFD) by Pablo et al. (2007).

2.1.1 Social Insects

The behavior of ants in social insect colonies is still an inspiring and most popular model in swarm intelligence. Wheeler (1911) described the social insect colony as an organism or as a higher-level organism or a super-organism because of the degree which the individuals appear to operate as a unit, coordinated function and physiology, dedicated to the perpetuation and reproduction of the colony as a whole. Wilson and Hölldobler (1988) further described the behavior of social insect colonies as an integrated unit that possesses the ability to process a large amount of information in a non-centralized manner. The colony members apparently make collective and individual decisions, e.g., task distribution, coordinate the activities of tens to thousands of workers, etc., and exhibits flexibility and robustness in response to external, i.e., environmental, challenges and internal perturbations.

Social insect colonies differ from each other in many ways. Some have only few individuals, whereas some include thousands or even millions of individuals. In some species, the individuals in a colony are short lived or seasonal. In others, they may live for many years (Queller and Strassmann 1998). In a colony, since the workers can perform a variety of tasks, different tasks are dynamically and distributively allocated to swarm or colony members, such as foraging, defense, nest construction, etc. Figure 2.1 shows one such example in which a group of ants collectively build a nest.

Fig. 2.1 Swarming: Oecophylla smaragdina workers are collaborating in nest construction, *source* http://upload.wikimedia.org/wikipedia/commons/a/aa/SSL11903p.jpg, by Sean.hoyland, used under license Wikimedia Commons/Desaturated from original, last accessed: 26 Aug. 2013

In literature different theories have been proposed to understand "how do the individuals in a colony or group interact with each other to develop a collective collaborative behavior?" "How does natural selection favor this kind of cooperation?". In his work of task allocation in social insect colonies, Gordon (1996) pointed out two kinds of factors that determine when and what task an individual worker performs: "internal" and "external" factors. The *internal factors* that are based on attributes of every individual in the colony are often considered to be fixed, i.e., body size, age and genetic factors. Whereas, the *external factors* are based on environmental stimuli and are therefore considered to be dynamic in nature. For example, when to go for foraging, a honeybee's decision might depend on how much nectar is already collected in the nest. Summarily, the actions of an individual are strongly influenced by at least two types of external cues: actions of other individuals and events happening in the colony's environment.

In the past couple of decades, different types of swarm or collective behaviors of different natural species were thoroughly investigated to solve or address a variety of problems in different domains, e.g., data mining, optimization, network routing, clustering, scheduling, traveling salesman problem, signal processing, etc. One characteristic feature of social insects, explored in recent years and linked with their foraging behavior is their ability to "organize" themselves in different groups or forms/shapes. This autonomous organization into multiple groups and subgroups involves mutual collaboration and collective efforts, that helps the swarm in several ways, e.g., dynamic task sharing or distribution, improved defense mechanism, better foraging, etc.

Self-organization is a process where a global structure emerges from the local interactions of a group of autonomous individuals without any centralized control mechanism. Bonabeau et al. (1999) defines four basic ingredients of self-organization: positive feedback, negative feedback, balance of exploitation and exploration, and multiple interactions. *Positive feedback* generally promotes the change by amplifying the fluctuations in the system, e.g., the intensity of pheromone laid by several individuals on a successful route. In contrast, *negative feedback* allows a system to tune its response in such a way that reduces changes. According to Garnier et al. (2007), self-organization requires multiple direct or indirect (stigmergy) local interactions among the swarm members to produce deterministic outcomes, and the appearance of large and viable structures.

2.1.2 Swarm Behaviors

Swarm behavior or intelligence is not just limited to social insect colonies. Rather, it can be seen in other species that show similar collaboration and cooperation in the environment, such as fish schools (Grúnbaum et al. 2005; Parrish et al. 2002), bird flocks (Reynolds 1987), sheep herds (Gautrais et al. 2007), or even crowds of human beings (Helbing et al. 2001). These species form or arrange themselves in groups or clusters with changing surroundings and environmental conditions, e.g., food supply, change of weather, etc. The clustering of these species for different purposes in nature evidently provide them several advantages, e.g., better navigation, improved defense against predators, improved or better foraging, etc.

- **Navigation**
  In nature, because of changing seasons, every day hundreds and thousands of birds and other animals migrate or travel from one region to another mainly for their survival, e.g., in search of better food, shelter, etc. The most fascinating aspect of their journey lies in their highly coordinated movement without any centralized control mechanism. By obeying simple principles, a flock without any collision travels several hundreds of miles in its everyday life. This collective and coordinated movement of individuals in different sized clusters help them to navigate more efficiently and effectively through different environments. The distributive but collective movement of bird flocks displays a structural order or pattern. The collective actions are so integrated that they appear to move as a single coherent entity, that has the tendency to abruptly change its shape and direction (Couzin et al. 2002). According to Grosan et al. (2006), the main principles of the collective behavior that is observed in a bird flock include:

  - *Homogeneity*
    All the agents (birds) of a swarm must be of similar species so that they behave identically.
  - *Locality*
    The movement pattern of an agent can only be influenced by its neighboring flock mates. For this purpose, an active vision among the swarm members acts as a sensor for the flock organization.
  - *Collision Avoidance*
    By maintaining a sufficient distance with its neighboring flock mates an agent usually avoids collisions in the flock.
  - *Velocity Matching*
    It is a critical parameter that needs to be dynamically regulated individually and collectively to exhibit a controlled coordinated movement in an aggregation,
  - *Flock Centering*
    It is a behavior that helps the flock members to stay close to their neighbors.

  With these simple rules the flock exhibit an extremely coordinated movement without any centralized control, which in result produces complex motion and interaction that would be extremely hard to create otherwise.

- **Defense**
  A significant advantage gained by individuals during swarming is the improved defense mechanism against the potential threats in the environment. For instance, a herd is less vulnerable to predators than an individual.
- **Foraging**
  "*Ants aren't smart, but ant colonies are*" (Gordon 2007). Foraging is among the best examples of decentralized distributed coordination of multiple individuals, found in nature. A colony with simple rules develops the tendency to collectively solve complex problems, such as, finding the optimum path to the best food source,

distributing tasks among colony members, etc.

Foraging is an act of looking or searching for food in the environment individually and collectively for the individuals and group survival. For this purpose, ants use an indirect way of communication mechanism to communicate with the colony members in the environment. While foraging, the forager ant lays pheromone trails to the food source to direct the fellow foraging ants towards the same food source. These pheromone trails later directly and indirectly help the foragers in finding the shortest path to a particular food source and the nest, and thus maximizes the net energy efficiency (Goss et al. 1989). The *pheromone* is a chemical substance, secreted and deposited by a foraging ant on its way to the food source back to the nest. These pheromone trails are then smelled by the fellow foraging ants or newly recruited foragers that then leave their own pheromone trails on their way from food source to the nest. This way the intensity of the pheromone helps the foraging ants to follow an optimum route to the food source and back to the nest. This serves as a positive feedback for the sharing of information about the available food sources in the environment (Sumpter and Beekman 2003). In Fig. 2.2 a group of foraging ants eating a slice of an apple is shown as an example of collective foraging. As already stated, to address the computational problems, e.g., optimal

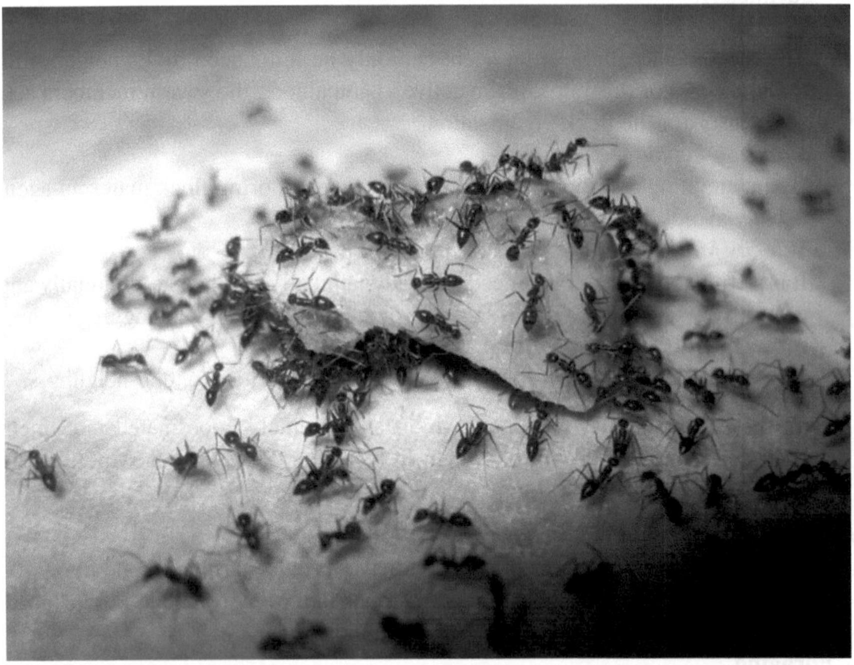

Fig. 2.2 Collective foraging behavior in an ant colony. Ants eating a slice of an apple, *source* http://upload.wikimedia.org/wikipedia/commons/c/ca/Ants_eating_fruit.jpg, *by* Zainichi Gaikokujin, used under Creative Commons Attribution 3.0 Unported/Desaturated from original, last accessed: 2 Nov. 2013

path finding, data mining, etc., the ant colony optimization algorithm is one such example that is inspired from the foraging phenomenon in natural ant colonies. Johnson (1982) categorizes the foraging strategies into two groups. The "solitary" and "group" foraging. The *solitary foraging* is a strategy in which the animals locate, capture, and consume their pray—food—on their own. It is normally seen in the scenarios where the food resources are in abundance. An example of an exclusive solitary forager is the South American species of the harvester ant, *Pogonomyrmex vermiculatus* (Torres-Contreras et al. 2007). In *group foraging* the animals locate, capture and consume food collectively with other individuals. Therefore, a failure or success in gaining a food resource depends on the mutual behavior of the foragers.

- **Reproduction**
  In honey bee colonies, swarming is the natural means of reproduction (Villa 2004). A new honey bee colony or beehive is formed when the queen bee leaves a colony with a large group of worker bees.

2.2 Swarm Robotics

In recent years, enormous research potential has been seen in mimicking the collaborative behavior of socialized animals into artificial systems. In the late 90's, Beni and Wang (1989) used a population of simple robots, called "cellular robots" to explore and develop the principles of swarm intelligence. Later, Kube and Zhang (1993) used the term "collective robotics" in their work of collective robot intelligence. During recent years, the term *swarm robotics* has emerged as an application of swarm intelligence to multi-robot systems that focuses on physical abilities of the autonomous individuals and their interactions among each other and between the individuals and the environment. Şahin (2005) defined the term as,

> *"Swarm robotics is the study of how a large number of relatively simple physically embodied agents can be designed such that a desired collective behavior emerges from the local interactions among the agents and between the agents and the environment."*

In general, "swarm" or "collective robotic systems" is a new approach to the coordination of fairly large population of relatively simple robots (Dorigo and Şahin 2004). The approach takes its inspiration from the system-level functioning of social insects—ants, termites, bees and wasps—which demonstrate three essential characteristics for a multi-robotic system: *robustness, adaptability, and scalability*, as mentioned earlier.

- **Robustness** in a system can be defined as the degree to which a system can withstand abrupt internal and external perturbations. In other words, it's an inherent ability of a system that allows it to continue operating in the presence of system failures or malfunctioning system components. For instance, the organization or distribution of workers in an ant colony continues operating even if they lost several members or if the environmental condition changes notably. The robustness in such a system can be attributed to several factors; the degree of redundancy in the system, that allows a swarm to tolerate the presence of faults (Kazadi et al. 2004). That is, at macro level, how a colony compensates the loss of its members, e.g., by recruiting new foragers upon the loss of existing ones. The decentralized coordination of individuals, i.e., the lack of any centralized entity or control mechanism, makes them invulnerable to potential break downs in the system (Melhuish 2001).

- **Adaptability** is a system's characteristic, either innate or acquired or both, that enables it to learn and adapt its critical system parameters that effect its stability in a dynamic environment. Adaptability and robustness have partly conflicting definitions (Bayindir and Şahin 2007). The difference between the two arises with the change in environment. For example, to withstand an unexpected change in the surroundings in which the system operates, it should be able to learn and adapt the change for its long term survival. Like in nature, animals change their behavior with the changing climate, e.g., food supply, temperature, shelter, etc., round the year. Adaptability in biological systems can be classified as: structural, behavioral and physiological adaptations (Bayindir and Şahin 2007). *Structural adaptations* point towards the physical appearance of the organism that aid its survival in its natural habitat, like, skin, shape, body covering, teeth, etc. The *behavioral adaptations* are linked with the behavior of an organism in a particular environment, e.g., search for food or a better shelter, etc. And, the *physiological adaptations* help an organism to adapt and regulate its internal system functions, like, secreting lime, body temperature, or like in ants, secreting pheromones.
  In the context of swarm robotics, it refers to the ability of individuals to modify or change their behavior over time, changes depending on the dynamic environment, changes in the assigned task, changes in the system composition or capabilities, in such a way that the net performance of the entire system can either be improved or at least not degraded (Iocchi et al. 2001).
- **Scalability** can be defined as "*the ability to expand a self-organized mechanism to support larger number of individuals without impacting performance considerably*" (Bayindir and Şahin 2007). In the context of swarm robotics, it can be defined as the collective ability of modules to maintain their performance and efficiency irrespective of the swarm population.

Considering the taxonomy of swarm robotics, Dudek et al. (1993) have proposed a classification of a *robot swarm* in terms of swarm size, communication range, communication topology, communication bandwidth, swarm reconfigurability and swarm unit processing ability. Later, Cao et al. (1997) segregated swarm robotics into: *group architecture*, *resource conflicts*, *origins of cooperation*, *learning* and *geometric problems*. Group architecture is then further divided into centralized and decentralized, homogeneous or heterogeneous robots, communication structure and modeling of other agents, dimensions. The modeling contains an individual's abilities, e.g., sensing, communication, locomotion, etc., to obtain or evolve effective and efficient cooperation in a robot swarm. Iocchi et al. (2001) also presented a taxonomy of multi-robot systems with the focus on the communication and computation aspects. In their classification four different levels are individuated in detail:

a *cooperation level*, a *knowledge level*, a *coordination level*, and an *organization level*. The knowledge level contains "aware" and "unaware" categories. The *aware* category at coordination level is then further subdivided into: "strong", "weak" and "non" coordinate categories. And finally, the strong coordinated category is further subdivided at organization level into: "strongly centralized", "weakly centralized" and "distributed" categories.

According to Bayindir and Şahin (2007) the mathematical modeling of swarm robotic systems is generally classified into *sensor-based, microscopic, macroscopic* and *cellular automata modeling*. The *sensor-based* modeling approach uses the models of the sensors and actuators of a robotic system along with the objects in the environment, e.g., recharge stations, obstacles, etc., as components of the modeled system. It further models the interactions between a robot and its teammates and with the environment. The sensor-based modeling approach is generally carried out using *non-physical* and *physical* simulations. The non-physical simulation does not include the physical properties (dynamics) of the robots and the objects in the environment (Hayes and Dormiani-Tabatabaei 2002; Howard et al. 2002a, b; Trianni et al. 2002, 2003). Whereas, the latter approach models the interactions between the robots and with the environment based on the physical rules, such as, assigning mass and force required to move or drag an object in the environment (Bahgeçi and Sahin 2005; Soysal and Sahin 2005; Trianni and Dorigo 2005; Trianni et al. 2006). The *microscopic modeling* approach captures the dynamics of a robotic module during its interaction with its swarm-mates and with the objects in the environment, but the details about its sub-components, e.g., sensors, actuators, body pose, are abstracted. The behavior of a robotic module is modeled using a state diagram and the transition between the states is triggered using internal (inside a robot), and external (environmental) stimuli. In literature, the microscopic modeling approach has been applied in a variety of applications. Likewise, Ijspeert et al. (2001), Martinoli and Easton (2003), Martinoli et al. (2004) used a probabilistic microscopic modeling approach to model the stick pulling problem in a robotic swarm. Lui (2008) used the probabilistic microscopic modeling approach to model the collective foraging task in a robotic swarm. The *macroscopic modeling* approach differs from microscopic models in terms of its granularity. The macroscopic modeling approach focuses on the behavioral aspects of the whole system, such as a robot swarm. The system behavior is usually defined in terms of difference equations, where each system state represents the average number of robotic modules in a particular state at that instance of time (Bayindir and Şahin 2007). The fourth kind of modeling approach is *cellular automata*. It is the simplest form of mathematical models of complex systems (Ilachinski 2001). It consists of a regular grid/lattice of cells where each cell has a finite number of states, such as *on* and *off*. Each cell interacts only with its neighboring cells based on some fixed rules—generally a mathematical function. The interaction determines the new state of each cell from its current state and the states of its neighboring cells. The studies conducted by Shen et al. (2004) in the context of self-organization and distributed control of robotic swarms can be seen as an example of cellular automata modeling. Figure 2.3 shows the taxonomy of swarm robotics in literature proposed by Bayindir and Şahin (2007).

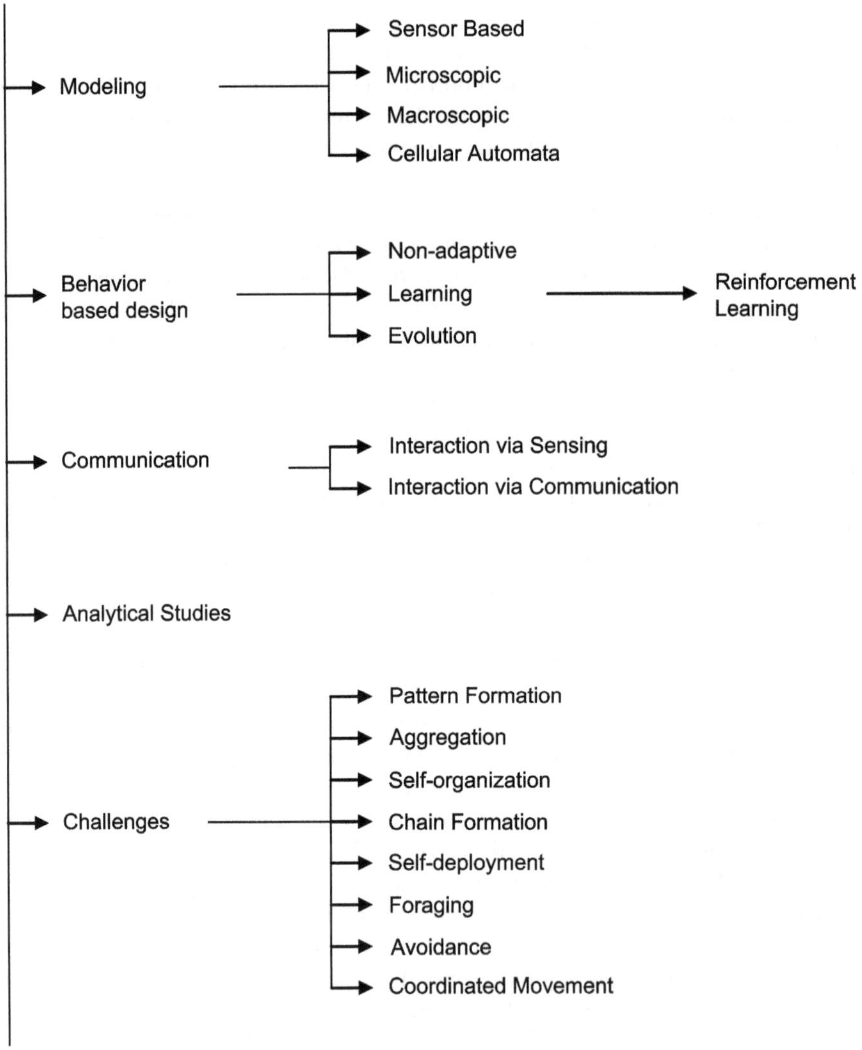

Fig. 2.3 Taxonomy of swarm robotics in literature (Bayindir and Şahin 2007)

The concept of modular robotics along with swarm intelligence principles has been explored in several applications using different platforms. These are broadly classified into two categories: *re-configurable with docking* and *re-configurable without docking*. The term "re-configurable with docking" capability is coined for those robotic platforms that own the ability to physically dock to each other either manually or autonomously and can adapt different morphologies for different purposes. The examples of such systems include PolyBot (Yim et al. 2000), SuperBot (Salemi et al. 2006), ODIN (Lyder et al. 2008), ATRON (Lund et al. 2005), Molecubes (Zykov et al.

2007), etc. Whereas, the second category of platforms though does own the modular structure but lacks the physical docking ability, for example, ePuck (Goncalves et al. 2009), Jasmine (Kornienko et al. 2005), etc. Hence, this category of robotic systems is only able to logically reconfigure themselves in different clusters.

In the context of this research work, the focus will be laid on different design parameters of only those re-configurable robotic systems that have the mechanical docking ability.

2.3 Modular Re-configurable Robotic Systems

The cells in a living creature may itself look quite simple but develop complex behaviors when aggregated in different morphologies—organisms (Fukuda and Nakagawa 1988). In recent years, creating simple robotic systems yet with the ability to adapt to the environment with complex morphing behavior gave birth to a new breed of robotic system known as *modular re-configurable robot*. A re-configurable robotic module with its innate ability of physical aggregation and dis-aggregation can dynamically adapt different morphologies. In the following section, two re-configurable robotic systems supporting some form of power management and energy sharing feature in their hardware design will be briefly described.

2.3.1 SuperBot

The SuperBot is a multi-functional, autonomous, re-configurable, homogeneous modular robot designed and developed by Salemi et al. (2006) at University of Southern California (USC). This modular robotic system was originally developed in collaboration with NASA for space exploration tasks. The salient features of a SuperBot robotic platform includes three degrees of freedom (DOF)—yaw, pitch, and roll—high-speed infra-red LEDs for communication, power sharing, re-configurable and flexible design enable the dynamic configuration of robotic modules to perform efficient locomotion and task manipulation. Figure 2.4a, b show a single SuperBot robotic module and a SuperBot modular robotic organism, respectively. The mechanical design of a SuperBot robotic module comprises two halves. One half of the robotic module contains a battery pack, a motor drive, a control mechanism for three docking sides, i.e., M1, M2, and M3, and a central controller called the "master controller", i.e., 8-bit 16MHz Atmega128 micro-controller. The master controller is connected to the controller of the second half, called the "slave controller".

The two controllers in every module are therefore responsible for managing the operations of actuators, communication links, sensors, power flow and the docking interfaces on each respective side. Figure 2.5 shows the block diagram of the hardware control architecture of a SuperBot robotic module. Each module has 6 docking

Fig. 2.4 Homogeneous SuperBot robotic system with a hybrid architecture (Salemi et al. 2006). **a** A single SuperBot robotic module. **b** SuperBot modular robotic organism

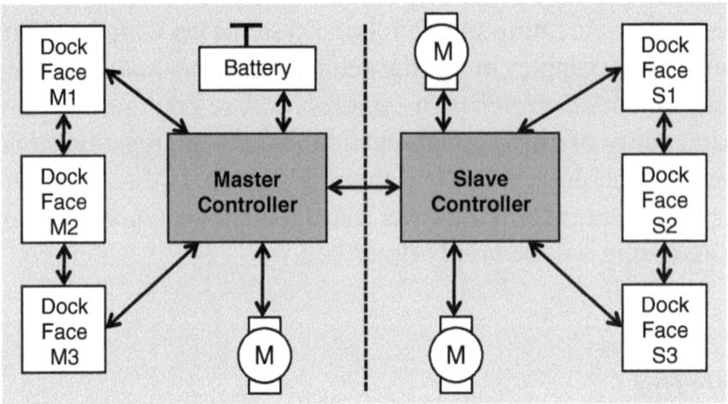

Fig. 2.5 Block diagram of the SuperBot hardware control architecture (Salemi et al. 2006)

connectors (sides) to allow the physical docking of other robotic modules in different configurations.

From the power management perspective, each robotic module is powered with two serially connected lithium polymer cells (2s1p) that provide a nominal voltage of 7.4 v with a charge capacity of 1600 mAh (Salemi et al. 2006). Figure 2.6 shows the SuperBot power sharing schematic. It includes 6 docking interfaces (3 on each half), and a recharging module. Each docking interface includes a diode that allows a uni-directional—inwards—current flow. Whereas, the outward current flow on each side is controlled by the controller with the help of a switch. Further, the local power bus interconnects the 6 docking interfaces thus allowing an omni-directional power flow in the system.

In the default system state the charging switch—the switch that connects the battery with the charging module—remains closed while the rest of the switches present on each docking interface remain open. The application of an external power source on any one of its docking interface, in the default configuration, allows the power to flow into the system's recharging module, thus allowing the on-board battery

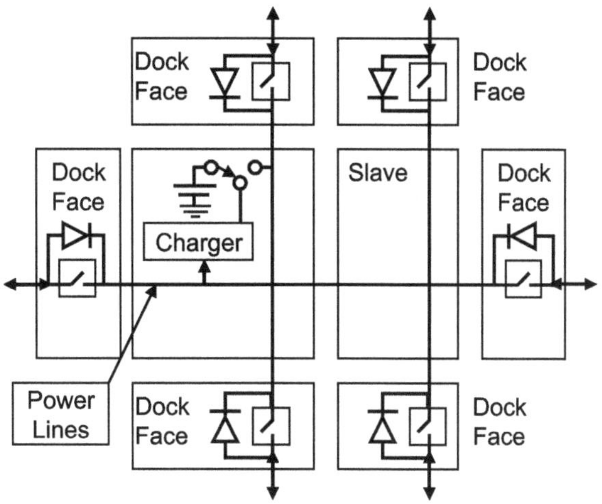

Fig. 2.6 SuperBot power sharing schematic (Salemi et al. 2006)

to recharge, instantly. This configuration of power bus switches in a modular robotic organism thus allows the turn-by-turn recharging of several robotic modules (in case, all the modules are dead) when any one of them is connected with an external power source. The power bus in a SuperBot modular robotic organism not only allows the recharging of robotic individuals through the external power source but also allows the modules to share their energy with each other in different configurations.

2.3.2 ATRON

The ATRON is a self-reconfigurable modular robot originally developed by the Adaptronics group at the University of Southern Denmark (USD) (Lund et al. 2005). Figure 2.7a shows an ATRON re-configurable robotic module. An ATRON robotic module consists of two hemispheres, joint together by a rotating mechanism that provides one DOF to the mechanical design. Each hemisphere has two passive female connectors and two active male connectors, which enable a module to hook to the complementary female connectors of the docking module (Østergaard 2004). Not only that, the two four-sided pyramids with craving allow the free rotation of robotic modules within an ATRON organism. Figure 2.7b shows an ATRON modular robotic organism constructed from a set of homogeneous, broadly spherical modules, with each of them 11 cm in diameter. The split design architecture, allows every ATRON module to rotate its one hemisphere with respect to the other to generate the collective movement patterns in the organism.

From the perspective of the system's electronic design, each hemisphere of the ATRON module has one 8-bit 16 MHz ATmega-128 micro-controller. Figure 2.8 shows the block diagram of electronics in an ATRON module. The northern hemisphere contains the main processor, an I/O processor, two docking connectors, a rota-

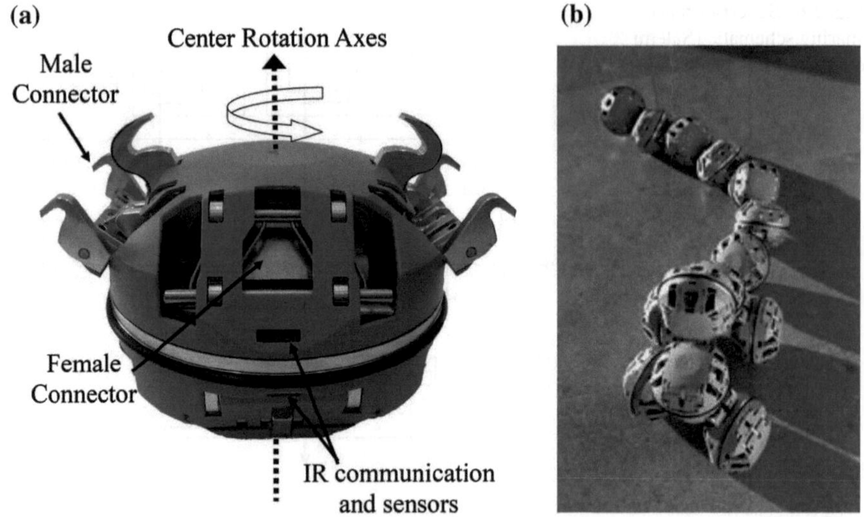

Fig. 2.7 Homogeneous ATRON robotic system (Østergaard 2004). **a** A single ATRON robotic module. **b** An ATRON re-configurable modular robotic organism

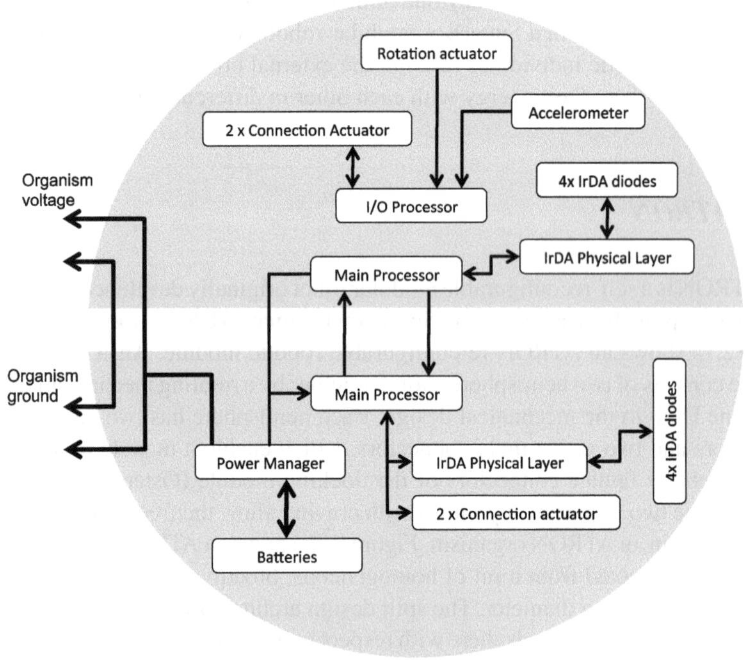

Fig. 2.8 Block diagram of the electronics in an ATRON module (Jørgensen et al. 2004)

tion actuator, an accelerometer, and four IrDA diodes. The main processor is responsible for communication with the northern hemisphere neighbors and the behavior of the entire module. The I/O processor, that is a 8-bit 1-MHz ATMega8 microcontroller, is responsible for reading the accelerometer, controlling the rotational mechanism, rotational lock, two active male connectors, communication and other system functionalities. The southern hemisphere in addition to the main processor includes a power manager, a battery pack, four IrDA communication diodes, and two active male connectors. The southern hemisphere's controller is responsible for intra-organism power sharing, recharging of the on-board battery pack, communication through four IrDA channels and controlling two active male connectors. Further, each ATRON robotic module is energized with two serially connected lithium-polymer cells (2S1P) that provide the nominal voltage of 7.4 V with a max. charge capacity of 980 mAh (Jørgensen et al. 2004). To maintain a consistent power supply to the robotic modules in an organism, the ATRON module's electrical skeleton enables the multiple modules to share or transfer their power with each other. This allows the robotic modules to recharge the neighboring module's battery pack. For this purpose, the entire supporting metal structure forming the skeleton of an ATRON module is electrically grounded. For applying positive electrical potential, one of the docking hooks has an electrically insulated piece of flexible printed circuit board glued on its top, that allows the module to form a power bus when physically connected (Jørgensen et al. 2004).

Figure 2.9 shows the power management system of an ATRON module. It includes a power manager, a charge manager, battery cells, and a CPU. The *power manager* in the system is mainly responsible for monitoring the on-board battery pack supply voltage and the voltage across the organism's power bus to select the best suited power source to power up its on-board electronics. The power sharing in a module is triggered if the organism's voltage (voltage across the power bus) is below a certain threshold with respect to the on-board battery pack's voltage. For collective recharging—parallel recharging of multiple modules in an organism—it is sufficient to connect a single module to an external power source. The modules then by election decide who gets to re-charge its battery pack first. A critical parameter,

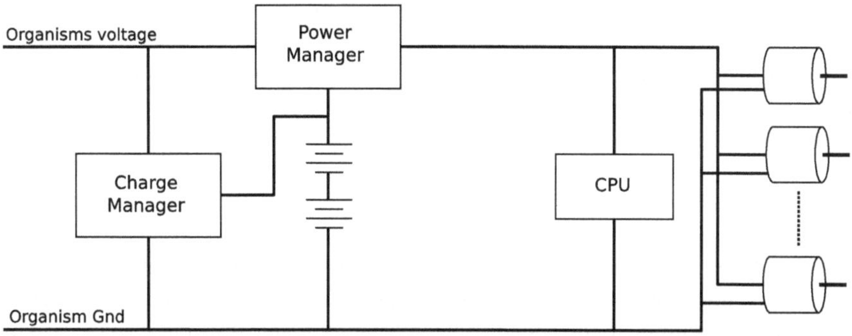

Fig. 2.9 ATRON module's power management system (Jørgensen et al. 2004)

that determines the maximum number of recharging modules in an organism at a time, is the maximum amount of instantaneous current that can flow through the organism's power bus. The ATRON module's system design allows a max. of 7 A of current to flow through. Therefore with 500 mA of recharging current, a maximum of 14 ATRON modules can be recharged simultaneously in an organism.

2.3.3 REPLICATORs

In a dynamic environment, the homogeneity of the robotic modules in a modular robotic organism may limit certain desirable features that are required, for instance, for collective locomotion on different surfaces, avoiding obstacles, lifting objects, exploring environment, etc. To gain such abilities, the idea behind heterogeneous platform design of REPLICATOR (Kernbach et al. 2008) robotic modules was intended to ensure high reliability, self-reconfigurability and adaptability in a modular robotic organism. The three heterogeneous robotic platforms specialized in different features are designed by the REPLICATOR consortium, namely; Kabot, Scout robot and Active wheel.

All three REPLICATOR platforms carry components that they have in common, e.g., docking units, cameras, transceivers for wireless and wired communication, IR sensors, microphones, accelerometers, and others. Among those components in which the three platforms differ due to their respective specialization belong the 3D-actuator, 2D-drive, and laser scanner. It is important to mention here, since the REPLICATOR robotic modules platform was partly designed and developed by the author therefore is used in this book to explore and address the issues related to the self-sufficiency of a modular robotic organism.

2.3.3.1 Kabot

The REPLICATOR *Kabot* is also called "backbone" robot because of its strong 3-D actuation. It was specially added as a feature for more robust, dynamic 3-D movement of robotic modules in the organism. Figure 2.10a shows the fourth generation of Kabot, with a 3-D actuation drive in the center. Its size is roughly $108 \times 108 \times 108\,\mathrm{mm}^3$.

The Kabot mechanical platform is specialized in strong 3D locomotion and actuation capabilities in the organism mode. The robot possesses a strong brushless drive capable of lifting several robots. A single DOF approach reduces the cost and since the main motor is a rotation-symmetric sub-assembly, the available space inside is better suited to support one strong motor instead of several smaller ones. To compensate the reduced number of DOFs, the frame of the robot uses two L-shaped halves, which can be rotated against each other. This allows lifting or rotating connected robots depending on their docked position. A symmetric, genderless, active docking was chosen to eliminate the need to search for compatible docking interfaces. This increases the diversity of possible organism structures that can be built. To use the

Fig. 2.10 REPLICATOR robotic modules (Kernbach et al. 2008). **a** Kabot. **b** Scout robot. **c** Active wheel

feature of the four mounted docking units, which not only provide for a stable connection but also for power and communication lines, a special 2D drive was included allowing the robot to additionally move sideways. The robot can move freely as a unit and it is possible to use the 2D drives of several connected robots to drive the organism. A cubical shape was chosen because it combines best the requirements for a small sized swarm robot and the requirements for modular self-reconfigurable robots, where a symmetric shape would simplify the reorganization of an organism.

2.3.3.2 Scout Robot

The REPLICATOR *Scout robot* is quite different in mechanical design from the Kabot. Figure 2.10b shows the fifth generation of REPLICATOR Scout robot. Its size is roughly $110 \times 110 \times 110\,\text{mm}^3$. Its platform is specialized in fast locomotion on challenging terrains for exploration tasks. Therefore, the platform is equipped with additional sensors compared to the other platforms to scan the surroundings and the floor. In particular, the Scout robot has two laser-camera systems, on its front and rear to provide far and short range obstacle detection. The Scout robot's locomotion is based on tracks. Fast locomotion for the exploration on rough terrains is more important for the Scout robot than slow and precise locomotion for aggregation and docking alignment, compared to the Kabot module. Tracks enable the Scout robot to move forward, backward, turn left, turn right and turn on its axis. Moreover, the Scout robot can move on rough terrains, climb slopes and overcome small obstacles. The speed of locomotion is more than one body length per second. With the continuous elastic tracks, the robot can perform locomotion even after overturning accidentally.

Similar to the Kabot, the Scout robot has four docking units, one on each lateral side wall. The docking units are centered on the walls of the robot so that modules can dock regardless of their orientation. This platform has 2 DOF, bending and rotation, with a maximum torque of 4,7 Nm. The bending DOF ($\pm 90°$) allows the robot's rear wall to lift up and down while the rotational DOF ($\pm 180°$) allows the docking unit on the robot's left side to rotate along its axis. The 3D DOF has lower torque than the Kabot since the main role of the platform is exploration and not macro-locomotion. In any case, the Scout robot can lift two other robotic modules. The Scout robot, with higher DOF and lower torque compared to the Kabot, performs best if docked to the tail or to the head of an organism to scan the environment.

2.3.3.3 Active Wheel

The *Active wheel* platform is optimized for specific tasks in the swarm and in the organism and is therefore initially regarded as an active tool in the REPLICATOR swarm. Figure 2.10c shows the third generation Active wheel. With its omni-directional and fast moving ability, it is specialized in transportation tasks. It is designed to carry and transport an organism consisting of several Scout or Kabots in a most energy-efficient way. The robot itself is able to approach the requested module or organism and can autonomously dock to it. The current design consists of two symmetrically arranged arms. On each end two 90° shifted omni-wheels are attached, which together form the locomotion mechanism of the robot. These arms are connected via a 180° turning hinge. Docking to other robots is provided by 2 docking elements placed on the same axis as the hinge. To allow docking in any position or height, the docking elements can rotate and are actuated by two dedicated motors.

The chassis of the robot has two cavities, which are ideal to place the main electronics and additional battery packs. On one hand, this configuration protects the sensitive components from potential damages during operations; and on the other hand, it provides enough room to place additional sensors for better perception of the environment.

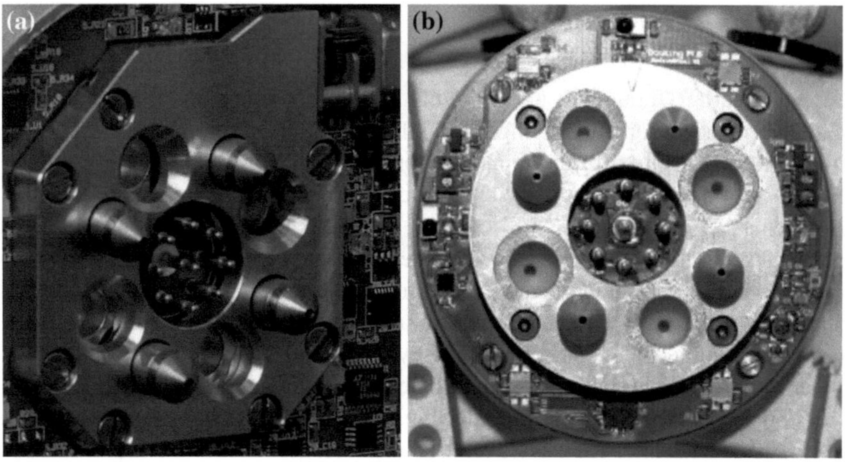

Fig. 2.11 REPLICATOR unified homogeneous docking unit. **a** Active docking unit on a lateral side of Kabot. **b** Passive docking unit on Active wheel

2.3.3.4 Homogeneous Docking Unit

A key element of re-configurable robot platforms is the ability to dock mechanically and electrically to merge into a larger artificial multi-cellular organism. A speciality of the docking unit developed for the REPLICATOR robotic modules, called CoBoLD—Cone Bolt Locking Device—is it's genderless design which is made 90° symmetric to handle slight misalignment between the robotic modules during an autonomous docking. Figure 2.11 shows the unified homogeneous docking unit. The design itself is based on 4 cone shaped, spring loaded bolts. Because of the self-locking feature of a worm gear, no energy is required to hold the connection. In the center of the docking unit, spring loaded electrical contacts are placed to assure an automatic connection of power and wired communication between the two docked robotic modules. The contacts are placed on the same PCB holding all the other electrical components on the particular side. This eases the electronic design since no cables or additional contact PCBs are required. The arrangement of these contacts plus a special switching circuit ensures that robots can dock no matter how they are oriented with respect to each other.

2.4 Bio-Inspired Techniques

This section describes those techniques that are inspired from the swarm behaviors of social insect colonies in nature, and applied in swarm robotics to gain energy autonomy and robustness. The bio-inspired techniques includes artificial foraging, trophallaxis, energy harvesting, homeostasis and fault tolerance.

2.4.1 Artificial Foraging

As stated earlier, foraging is not only vital for the life of any living system but also equally important for a non-living system that tries to keep autonomy. *Artificial foraging* (AF) is inspired from the natural foraging behavior found in social insects. It is an act that enables an artificial robotic system to search and regain its replenished energy (food) from the environment. The first known robot that possessed the artificial foraging ability was believed to be the "Tortoise" robot developed by Grey (1963). Here, the foraging behavior was replicated using a simple light following approach that allowed it to guide itself towards the recharge station in the arena. AF behavior is quite analogous to the "basic cycle of work" defined by McFarland and Spier (1997), to maintain the "self-sufficiency" of an autonomous mobile system (McFarland 1994). Foraging and collecting behavior in artificial autonomous robots is replicated in a variety of ways. For instance, a forager robot or a group of foraging robots (Goldberg and Matarić 2000; Parker 1995) has to collect objects scattered in the arena and to bring them collectively or individually either in some random location called the *collection task* (Beckers et al. 1994; Matinoli et al. 1999), or a pre-specified location termed "home" known as the *foraging task* (Goldberg and Matarić 2000; Matarić 1992; Nitz et al. 1993). During the past two decades, several algorithms have been proposed that explore the foraging behavior in robotic systems. The first known implementation of foraging task in a group of real robots was done by Matarić and Marjanovic (1993), Matarić et al. (1995). Figure 2.12 shows Matarić's robots used for foraging tasks. The proposed foraging behavior defines the following states: *avoiding*, *dispersing*, *searching*, *homing*, and *resting*, which are called behavior primitives.

Fig. 2.12 Matarić's foraging robots (from Matarić 1997)

In biological systems, the *efficiency* of foraging behavior has been defined in terms of "energy" that is measured as the weight of animals before and after foraging to quantify the amount of energy spent and gained (Charnov 1976; Houston et al. 1988; Oster and Wilson 1979). In the context of swarm robotics, the terms such as, *reward, income, benefit* has been used by Labella et al. (2004), Lerman et al. (2006), Lui et al. (2006), Ulam and Balch (2004).

Lui (2008) in his Ph.D. work modeled a collective foraging scenario in which the food-items were randomly scattered in the arena. A swarm of robots was assigned the task to collect and deposit the food items to a pre-specified location, called *nest*. Each food item that is collected by a robot provides an amount of energy to the swarm, but the foraging activity requires a certain amount of energy for searching and grabbing a food item. The optimal goal of the developed swarm was to maximize the net energy of the swarm.

The collective foraging behavior was modeled using a finite state machine (FSM), shown in Fig. 2.13. Each state in the finite state machine represents a sub-task that a robot undergoes during foraging. The transitions between the states are triggered on the basis of internal, i.e., time, or external, i.e., sensors, events occurred to a robot. The foraging FSM uses two types of timers, i.e., T_s and T_r, to count the time spent by a robot during, *searching* and *resting* in the nest. The transitions from states *randomwalk, scanarena, or movetofood* to state *homing* are triggered whenever *searching time* T_s reaches the threshold Th_s, i.e., $T_s \geq Th_s$, that ultimately reduces the number of foraging robots to resting robots. The transition between state *resting* and state *leavinghome*, is triggered when the resting time T_r reaches the threshold Th_r, i.e., $T_r \geq Th_r$, thus drives the robot back to foraging. The proposed adaptation that is biologically inspired uses three kinds of cues: *environmental cues, internal cues and social cues*, to adjust the time threshold parameters Th_s and Th_r that dynamically change the number of foraging and resting robots in the arena to improve the efficiency of the swarm. The performance metric of the swarm was calculated as the energy efficiency given by Eq. 2.1,

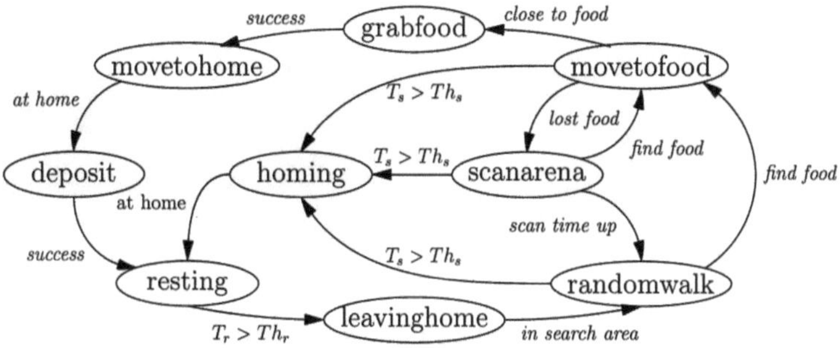

Fig. 2.13 Finite state machine for collective foraging task (from the work of Lui 2008)

$$\text{efficiency} = \frac{\text{net energy income to swarm}}{\text{energy available from environment}}. \quad (2.1)$$

Where the term in the nominator, i.e., *net energy income to swarm*, is the amount of energy gained every time a food item is collected by a forager from the arena minus the energy spent or invested to collect that food item. And, the term in the denominator, *energy available from environment*, is the total number of food items growing over time multiplied by the energy unit a food item provides.

2.4.2 Trophallaxis

"*In the social Hymenoptera, continued evolution has placed a premium on the storage of nutrients for the benefit of the society as a whole. They have two central chambers in their digestive tract —the crop—followed by the midgut*" (Eisner and Brown 1956). The crop holds the food that can be exchanged between the foraging and non-foraging nest-mates and is therefore called the ant's *social stomach*. The exchange of food between the individuals of a colony serves as one of the most fundamental bond in the social organization of ants (Eisner and Wilson 1958; Wheeler 1923; Wilson and Eisner 1957).

Trophallaxis in biological systems is a process that allows conspecifics—two or more individual organisms of the same species—to exchange or transfer food through regurgitation via "mouth-to-mouth" known as, *stomodeal* (Korst and Velthuis 1992) and "anus-to-mouth" known as *proctodeal* (Cabrera and Rust 1999). Figure 2.14a shows food transfer—trophallaxis—between two ants. Figure 2.14b shows the trophallaxis behavior between two honeybees in the hive. The sharing of food between the colony members is not limited to feeding the individuals rather it also provides a means of communication (Camazine 1993; Korst and Velthuis 1992), like in honeybee colonies (Wainselboim and Farina 2003). In addition, it also boosts colony-level immunocompetence, known as "prophylaxis" (Hamilton et al. 2010).

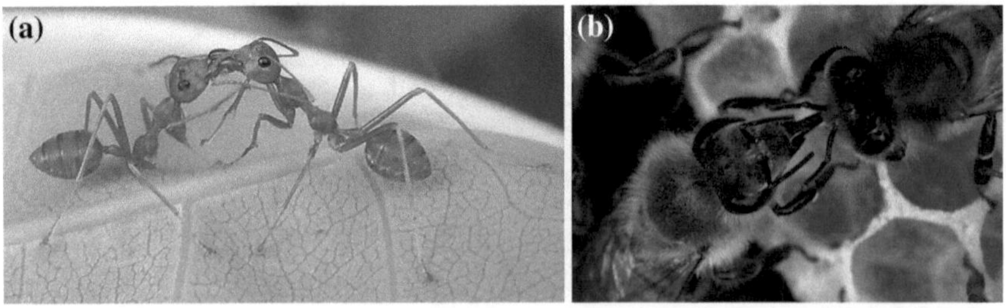

Fig. 2.14 Trophallaxis behavior in nature. **a** Ants engaged in mouth-to-mouth food transfer (*source* http://upload.wikimedia.org/wikipedia/commons/d/dd/SSL12022p.jpg, *by* Sean.hoyland, used under license Wikimedia Commons/Desaturated from original). **b** Trophallaxis between two honey bees—(mouth-to-mouth) food transfer (with the kind permission of Eric Tourneret, author)

In swarm robotics, the term *trophallaxis* has been applied for both energy homeostasis in a swarm (Melhuish and Kubo 2007) and as a means of non-centralized communication within a large robot swarm colony as presented by Schmickl and Crailsheim (2008). Trophallaxis is specifically found useful in insect colonies that need to regulate the internal state of the colony, e.g., the protein supply in the bee hive, without having any centralized control, such as, a brain and communication path-way, like, neurons. Hence, for swarms of disjoint robots it can similarly aid in regulation and self-organization. A re-configurable robot however, does have a dedicated communication channel and may in some settings also provide a central "brain" by using distributed processing in only a few of specialized modules. It is even questionable if the term trophallaxis is valid at all *within* a robotic system consisting of merged[1] robotic modules that simply share their energy on a common energy bus. Trophallaxis may consequently not have the same significance in re-configurable robots, unless these themselves are a member of a swarm of *organisms*. Generally, in a swarm of self-reconfigurable robots one will find single, disjoint robot modules and robot organisms of different morphologies consisting of a variable number of merged modules. If two such robot units, single or merged, dock together for the sake of energy transfer, the term trophallaxis is regard as being a valid analogy.

Ngo (2008) in his work, "towards sociable robots", presented the concept of artificial trophallaxis in swarm robotics. For that, he developed a new robotic platform named "CISSbot" (Ngo and Schiøler 2008). The CISSbot has the physical dimensions of 15×15 cm square shape. The mechanical platform is driven by a differential two wheeled system and a pair of castors to maintain the robot balance.

Trophallaxis based on recharging the other's battery is quite time-consuming (considering today's battery technology up to several hours) and inefficient from an energetic perspective. Hence, Ngo (2008) has devised mobile swarm robots that exchange energy through swapping of battery packs. The battery packs are carried on their top out of a specifically designed battery holster unit that can hold up to 8 ejectable batteries, as shown in Fig. 2.15. The idea behind this was to increase the sustainability of the swarm, since the individual robots do not have to interrupt their ongoing tasks for re-fueling through a recharge station. Figure 2.15 shows the artificial energy trophallaxis—swapping of battery packs—between two CISSbots.

In contrast to biological systems, it is important to emphasize that by trophallaxis a swarm of robots does not gain energy. Rather, a significant portion of energy is lost, since the efficiency of energy transfer will always be less than perfect. The energy loss during the exchange is usually not as critical as it is in artificial systems. However, a swarm may prolong its autonomy through trophallaxis, by taking over the role of a "common stomach", levitating any peaks and dips of energy supply and assuring a more evenly distribution of power supply (Humza and Scholz 2010).

---

[1] Although the robot hardware cannot merge, this term is being used but to underline that the modules do not only physically aggregate and bond, but also unifies logically—they become one larger entity.

Fig. 2.15 Artificial energy trophallaxis: two CISSbots are in a state of exchanging batteries (with the kind permission of Ngo) (2008)

2.4.3 Energy Harvesting

Robots that move and operate autonomously must have access to some external energy resources to maintain their autonomy. Although in recent years considerable progress has been made regarding the output power capability and the energy weight ratio of batteries (Linden and Reddy 2002), but the stored energy is still very limited compared to the power and energy demand of a meso-scale robot module. As is easily imaginable, this shortage becomes worse when energy-hungry, highly sophisticated sensors and processing systems are involved such as vision systems. In particular for self-reconfigurable robots, it is expected that in general a higher demand of sensing and processing power is required compared to "conventional" swarm robots that only move in 2D space. To increase the energetic autonomy it is hence wise to enable the robots to collect additional energy from the environment, which is often called *energy harvesting*.

Depending on a robotic module's architecture and its operational environment, different forms of energy harvesting methods have been employed, like solar power (Boletis et al. 2006; Landis and Jenkins 1997), wind power, vibrational power (Wade and Gifford 2007), etc. The robotic systems that employ energy harvesting methods based on fueling from organic substances include: Gastronome that uses sugar (Wilkinson 2000), Ecobot-II that extracts power from sludges and flies (Ieropoulos et al. 2005b), and slugs eater using microbial fuel cells (MFC) (Kelly et al. 2000). Figure 2.16a shows a Gastronome robot and b shows an Ecobot-II that uses organic substances to fulfill its power demand. A microbial fuel cell that performed best, provides an average output power of 45.5 μW over a period of 10 days and has the dimensions of $6 \times 7 \times 5$ cm (Ieropoulos et al. 2005a).

When considering the power demands of a robot, its size is an extremely important factor since technologies for energy storage and harvesting do not scale down that easily. For example, if a cube shaped robotic module is shrunk in length by a factor x, the surface area would decrease by x^2, whereas its volume would be reduced

Fig. 2.16 Energy harvesting robots. **a** Gastronome—a prototype MFC powered robot, copied from Wilkinson (2000). **b** An Ecobot-II with O_2 MFCs onboard (Ieropoulos et al. 2003)

by x^3. In a rough model, the energy consumption due to the weight of the robotic module is more or less proportional to its volume. On the other hand, solar cells generate electrical energy approximately proportionally to their surface area. This implies that it will be much easier to power a small robot with solar cells than a larger one. Reports on energy scavengers confirm the scaling issue. Even though energy harvesting is being employed in more and more marketed mobile electronic devices (vibrational energy scavenging in wristwatches, thermal energy transformers for wearable sensors, etc.) many of these are not very useful for robotic applications in the meso-scale. This conclusion is drawn from the fact that the reported appliances rely on extremely low power demand electronics and lack any actuator, a key component in robotics that due to physical limitations cannot be made energy preserving to an arbitrary level. It is certainly not possible to give exact numbers of robots' power consumption in general. There are several factors that come into play, such as size and hence the weight, processing power and grade of sophistication, principle of locomotion, number of sensors, and many more. In Mei et al. (2005) the author has tried to shed some light onto mobile robots at meso-scale. When looking at the given numbers (10–20 W) it is obvious that certain energy harvesting methods at the current state of art are not able to provide the necessary power. When considering that self-reconfigurable robots will usually be more complex than ordinary swarm robots (sensor fusion for docking approach, docking mechanism, etc.) energy scavenging by integrating harvesting modules within the robot modules is even less promising because of the tighter size restrictions and the expected higher energy demand due to 3D actuation, when lifting other modules.

Nonetheless, photo-voltaic cells are commonly being used for robot energy scavenging, in particular in space missions. The efficiency of solar cells not only depends on the specific technology being used (Si, Ga, thin film, crystalline, etc.) but also on the spectrum of light. Hence, the cells are compared at a well-defined spectrum, which has been standardized, e.g., IEC60904-3, at 1000 W/m^2. This irradiance prevails approximately at a cloudless noon during spring in central Europe. The highest reported efficiency of PV modules—these are composed of several cells connected together—lie between 8.2 % (thin-film poly-crystalline Si) and 22.9 % (crystalline

Si) (Green et al. 2009). Consequently, a solar panel made of crystalline Si with a realistic efficiency of 15 % under ideal conditions will generate approx. 150 W per square meter on an average bright sunny day. To power a re-configurable robot this may still not be enough, when taking a much smaller available surface area into account. But for a base or charging station to which the robot modules return to recharge this may be a realistic scenario.

2.4.4 Homeostasis

The biological term *homeostasis* or *milieu interieur* was first coined by the American physiologist Walter Bradford Cannon (Cannon 1926, 1929). Later, the following definition of homeostasis comes from his book "*The Wisdom of the Body*" (Cannon 1932):

> "*The coordinated physiological processes that maintain most of the steady states in the organism are so complex and so peculiar to living beings—involving, as they may, the brain and nerves, the heart, lungs, kidneys and spleen, all working cooperatively—that I have suggested a special designation for these states, homeostasis. It means a condition—a condition that may vary—but is relatively constant.*"

Claude Bernard, a French physiologist, was one of the first to discuss what is now called homeostasis, frequently talking about the concept in terms of the stability of the internal environment (*milieu interieur*). Bernard and Dastre (1879) states:

> "*stability of environment implies an organism so perfect that it can continually compensate for and counter balance external variations. Consequently, far from the higher animals being indifferent to their surroundings, they are on the contrary in close and intimate relation to them, so that their equilibrium is the result of compensation established as continually and as exactly as if by a very sensitive balance.*"

Wheeler (1911) described the regulatory behavior of social insects colonies as a "social homeostasis". *Social homeostasis* is a phenomenon, in which the collective activities of a colony members help them to regulate the colony environment. For example, the regulation of nest atmosphere in termite colonies: the concentrations of carbon dioxide, oxygen, and water (Emerson 1956), respiration in honeybee colony, etc.

In a physiological context, the word "*homeostasis*" describes a property of a living system that is able to maintain a stable internal environment, despite the potentially considerable variations to its internal and external conditions. The *steady states* that Cannon referred to are the relatively constant values at which physiological variables such as blood pressure, body temperature and blood glucose levels are kept, what one might call "organismal homeostasis". As Cannon states, the stability of these variables is maintained by a number of "coordinated physiological processes". One such example is the *thermoregulation* of the human body. During thermoregulation, the variable to be maintained is body temperature, which is kept at a near constant 37 °C, despite fluctuations in the external temperature. A classic example of a stable engineered system is the Watt's governor, a type of centrifugal governor designed in 1788 by James Watts (1736–1819) to regulate the speed of a steam engine. The Watt governor, and other engineered systems involving more state-of-the-art proportional control mechanisms are very good at maintaining the stability within the small task they were designed for, however, they lack the adaptability that stable biological systems possess. Biological homeostatic systems, as has already been shown, are highly adaptive.

Here, it is important to introduce the terminologies required to define homeostatic behavior, later used to define the homeostasis from the perspective of self-sufficiency of a modular robotic organism. Following terminologies largely originate from the two sources: Widmaier et al. (2006)'s book "*Vander's Human Physiology: The Mechanisms of Body Function*" on physiology and Ashby (1960)'s book "*The origin of adaptive behavior*".

- **Essential variables**
  It refers to those variables that belong to the internal state of an organism and are closely linked with its survival. The deviation of an essential variable from its normal range will lead to changes in the organism that compromise its survival. Examples of essential variables include: body temperature, blood pressure and blood glucose levels, etc.
- **Set points**
  It refers to those values that define the system steady state, around which each essential variable in a healthy organism will tend to remain. Essential variables do not remain constantly fixed at their set point values. In a homeostatic control system, when a variable deviates from its set point, several physiological processes will come into action that tend to bring the variable back to its steady state. For instance, the set point of the human body temperature is 37 °C. It should be noted, the set points are not fixed for the life-time, rather they are adaptable, to accommodate the changes happening in the operating environment. For instance, the human body temperature becomes higher during the day when the body is active.
- **Physiological limits**
  They define the normal range of an essential variable. If a variable travels outside its limits then the survival of the organism is put at risk. For example, the physiological limits of the human body temperature are considered to be approximately 35–40 °C, an internal temperature of below 35 °C may lead to hypothermia whereas an individual with a temperature above 40 °C may experience heat stroke, leading to hyper-thermia.
- **Negative feedback**
  It is the property of a system in which the increase or decrease in the value of a variable initiates the responses that tend to oppose the change to that variable. Within the endocrine system, for example, a negative feedback loop controls the release of many hormones in the human body. The increase in the concentration of a particular hormone often leads to a series of responses that in the end tend to inhibit or prevent the release of that hormone.

- **Regulatory response**
  In response to a stimulus it is a physiological process that acts so as to maintain the essential variables within their physiological limits. Regulatory responses often form part of a negative feedback loop. For example, in response to a drop in body temperature the regulatory response of shivering will bring about changes that tend to increase body temperature. Responses can either be *involuntary, unlearned, built-in*, or *acquired* and subject to learning and improvement over time. According to Widmaier et al. (2006), since *built-in responses* are subject to learning therefore it is often hard to distinguish the following two types.
  - **Feed-forward regulation**
    It is a special form of regulatory response that compensates for changes to *essential variables* before any such change has actually occurred. Feed-forward regulation further minimizes the time and the extent which the variables are at risk of breaching their physiological limits.
  - **Acclimatization**
    Improvement of an already existing homeostatic system due to prolonged exposure to certain environmental conditions. The adaptation of the sweating response in different climates is a good example of acclimatization. The sweating response of an individual that is used to exercising in a hot environment will start quicker and act more aggressively than the response of an individual that is acclimatized to a colder environment.

2.4.5 Fault Tolerance

Tolerance to the presence of faults is an important consideration which an autonomous system must take in to account for its long-term survival. *Fault tolerance* is the ability of a system to continue its operation in the presence of faults and component failures perhaps by gracefully degrading its system performance (Parker 2012). It has been explored in a variety of application scenarios, and attained through different means. In artificial control systems, it is usually addressed through hardware redundancy and redundant control behavior. *Hardware redundancy* or *replication of hardware* is most commonly employed to enhance the system reliability by physically duplicating the critical system components or modules in the system design, that increases the endurance to faults and failures (Johnson 1984; Siewiorek and Johnson 1981). Later, Payton et al. (1992) had investigated a set of redundant control strategies to address high level failures, to gain a high level control and fault tolerance. A *high level failure* refer to those malfunctions that a system can encounter for which it is not explicitly pre-programmed. To deal with unexpected system behaviors, the controller is usually designed with redundant control strategies to perform a task. For instance, if a strategy does not suffice, the controller selects or switches to one of the other defined strategies (Ferrell 1993). Considering the two strategies, the hardware replication strategy is not acceptable to every system design as the redundant system components not only increase the size of the system but also adds complexity and development cost. On the other side, the redundant control behavior based strategies to encounter faults does not suited well for hardware faults and failures.

In the context of modular robotic systems, the increased system complexity may in some scenarios result in higher failure probability. Carlson and Murphy (2005) report a wide variety of reasons why individual robots fail in the field. According to them, the different causes of failure can be roughly categorized as physical failures, software control failure, or human control errors. Winfield and Nembrini (2006) use the approach of qualitative Failure Mode and Effect Analysis (FMEA) (Dailey 2004) to evaluate the fault tolerance of a robot swarm. Under the FMEA approach, the designer tries to identify different sources of hazards that can cause *internal* and *external* failures in a robotic system. For each hazard, an analysis is performed to determine its impact on the performance of a robot swarm. The tolerance ability mainly depends on the detection, identification and isolation mechanism that allows a system to operate in the presence of malfunctioning components without the need of external intervention. *Fault detection* is a complex process for a number of reasons as pointed out by Parker (2012): the space of possible fault types, sensors, actuators, and the uncertainty in the environment models. *Fault detectors* are commonly employed to monitor a system's essential variables. Willsky (1976) presented a survey of design methods for failure detection in dynamic systems. Upon detection of an unacceptable deviation of a system's parameter, a fault is diagnosed. The *fault diagnosis* locates the fault and establishes its cause to evaluate its effect on the system steady state operations. If the evaluated fault is tolerable, the system may continue its operations; otherwise, the system adapts its behavior accordingly, i.e., through fault isolation and accommodation (Isermann 1984; Paul 1990).

Fault detection and isolation (FDI) is based on the use of *analytical* instead of *physical* redundancy (Paul 1990). It exploits the available information redundancy in the system (Garcia et al. 2000; Jeppesen and Cebon 2004; Leuschen et al. 2002). FDI techniques can be broadly classified into: *model based* and *knowledge based* models. Different model based approaches for fault detection have been developed over the years, see, e.g., (Gertler 1998; Himmelblau 1978; Isermann 1984, 2005; Isermann and Ballé 1997; Paul 1990; Willsky 1976). For knowledge based FDI approaches, see, Milne (1987) and Patton et al. (1989). The concept of *analytical redundancy* is commonly employed for *failure detection* and *isolation* in many systems. Without employing redundant components, it evaluates the inherent system redundancy through information processing under well-featured operating conditions. Figure 2.17 shows the general architecture of the FDI approach based on analytical redundancy, redrawn form Isermann (1984).

The analytical redundancy approach employs residual generation for the detection of faults in the system. The residual generation requires the system models during the three operating modes: *nominal*, or *steady state*, *observed*—actual—and *faulty system*. In the *fault decision logic* the decision on the occurrence of a fault is usually based on the *fault signature* generated from the faulty system model. On the detection of a fault its location, size and cause can be determined using fault diagnosis or inter-

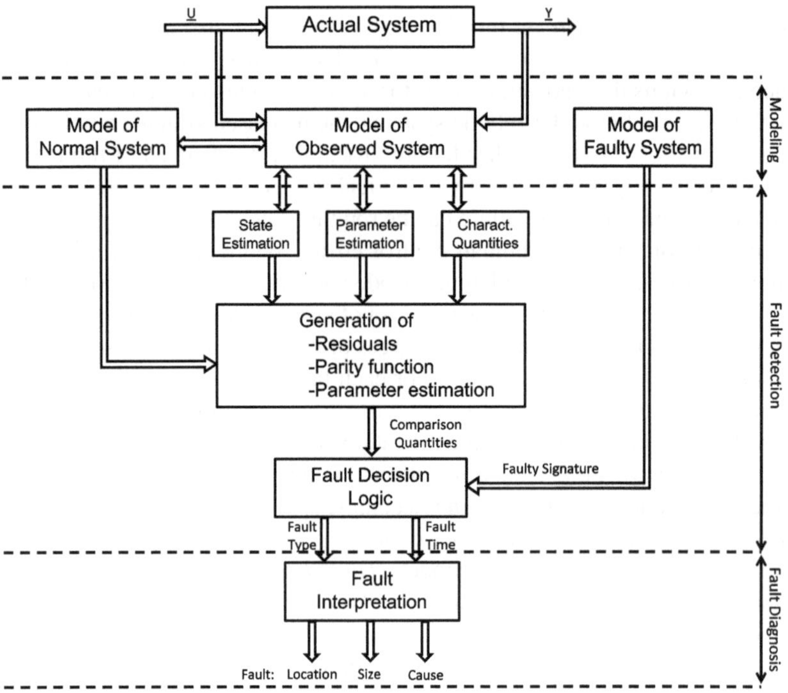

Fig. 2.17 General architecture of FDI based on analytical redundancy, redrawn from Isermann (1984)

pretation. For this a deeper knowledge about the system, e.g., aging, the operational environment, used tools, fault statistics, etc. is required (Paul 1990).

To capture the dynamics of an artificial system especially for the construction of a *faulty system model*, it is important to firstly, distinguish between failure mode and failure mechanism (Larsen 1974). A *failure mode* is a type of *failure* that causes deviations in the essential variables of a homeostatic control system, while, *failure mechanism* indicates the cause of a particular *failure* in the system. For this purpose, H.A. Watson in 1962 at Bell Telephone laboratories first used the *fault tree analysis* approach to analyze the Minuteman Launch Control System (Driessen 1970). *Fault tree analysis* (FTA) is a top-down approach to failure analysis. It identifies the various sequence of events that lead to a component malfunctioning. It starts with a potential undesirable event (failure) as a TOP event and then traces down all the way to its root cause or causes. A fault tree is a logic diagram based on statements that are either true or false, on or off, open or close, etc. (Larsen 1974). An event in the fault tree may either represent a component failure, an external influence/disturbance, or a system operation. The events in a fault tree are connected by a set of logic symbols. Figure 2.18 shows the fault tree analysis symbols.

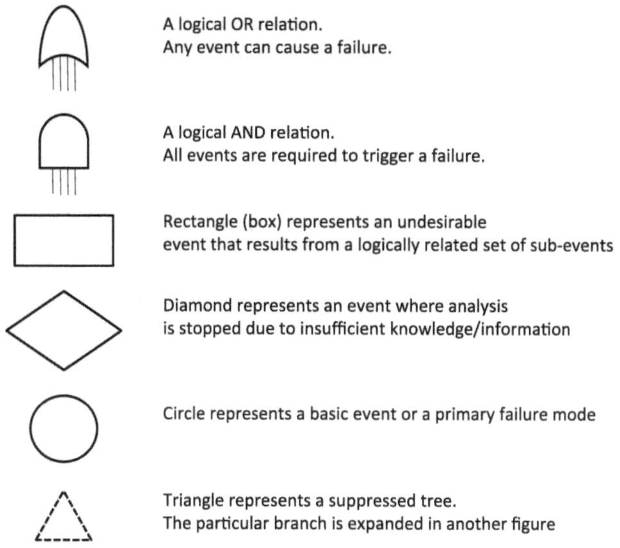

Fig. 2.18 Fault tree analysis symbols (from Larsen 1974)

2.5 Simulation Tools

2.5.1 Player/Stage

The open source project Player/Stage (Gerkey et al. 2003), provides an option for research into robotics and sensor systems. Player is probably one of the most widely used open source robot interfaces in research and post-secondary education (Toby and Bruce 2005). The components of the simulation framework include the Player network server and Stage robot platform simulators.

2.5.1.1 Stage

Stage simulates a population of mobile robots, sensors and objects in a two-dimensional bit mapped environment. It provides fairly simple, computationally cheap models of many devices rather than attempting to emulate any device with great fidelity. This is particularly useful for simulation of a large group of mobile robots. It is often used as a Player plugin module, providing populations of virtual devices for Player. Users write robot controllers and sensor algorithms as 'clients' to the Player 'server'. Typically, clients cannot tell the difference between the real robot devices and their simulated Stage counterparts. After careful design of the simulated robot model in Stage, Player clients developed using Stage will work with little or no modification with the real robots and vice versa. Thus Stage allows rapid prototyping of controllers destined for real robots. Various sensors and actuators are provided, including sonar, scanning laser range finders, vision (color blob detection), odometer, and a differential steer robot base.

2.5.2 *SPICE*

SPICE (Simulation Program with Integrated Circuit Emphasis) is a simulation and modeling program for analogue circuits and digital logic design. It is used in integrated circuit and board-level design to check the integrity of circuit designs and to mathematically predict the electronics circuits behavior. It provides an environment to simulate components ranging from the most basic passive elements such as resistors and capacitors to sophisticated semiconductor devices such as MESFETs and MOSFETs. Using these intrinsic components as the basic building blocks for larger models, designers and chip manufacturers have been able to define a truly vast and diverse number of SPICE models. Most commercially available simulators include more than 15,000 different components. An electronic circuit in SPICE is defined in the form of a net-list. The net-list is a text description of all circuit elements such as transistors and capacitors, and their corresponding inter-connections. Different schematic capture and simulation tools such as Multisim from National Instruments, LTSpice from Linear Technology, etc., provide a graphical user interface to draw circuit schematics in a user-friendly environment, and automatically translate the circuit diagrams into net-lists.

For the simulation of dynamic power flow and fault tolerance between the robotic modules of modular robotic organisms, in this research work, LTSpice has been used to predict the system behavior at the hardware layer. LTSpice IV is a high performance SPICE simulator, schematic capture and waveform viewer with enhancements and models for easing the simulation of switching regulators. It provides macro models for 80 % of Linear Technology's switching regulators, over 200 op-amp models, as well as resistors, transistors, and MOSFET models. Chapter 4 introduces the simulation framework that uses LTSpice with a simulation front-end designed in LabWindows/CVI to simulate power flow in different organism morphologies for this work.

2.6 Summary

This chapter briefly covered the background and the related work carried out in autonomous robotic systems, mainly from the perspective of self-sufficiency. It started with the basic definition of swarm intelligence and its five basic properties. It discussed the swarm behaviors observed in nature in different species that form the basis of swarm intelligence. The common examples of swarm behaviors include navigation, defense and foraging. The chapter then focused on swarm robotics by outlining three essential characteristics of a multi-robotic system: robustness, adaptability, and scalability. After including the taxonomy of swarm robotics in literature, it briefly discussed the platform design and features of some of the re-configurable modular robotic systems that support power sharing. After having a detailed analysis of the state-of-the-art re-configurable modular robotic systems, the following features may be found to be missing in the existing systems:

- determine or measure the current flow through the on-board battery pack to system peripherals and to each docking side—to the docked robotic modules in the organism,
- withstand abrupt system failures, such as, a short circuit, erroneous sensor measurements, component failure, etc.,
- control the recharging of an on-board battery pack of a robotic module in organism mode, especially in case of a faulty robotic module,
- application layer control mechanisms for dynamic power sharing between the robotic modules of an organism.

The current flow measurements at the critical system segments play an important role in controlling the dynamic operations of any battery operated autonomous mobile robotic system. These measurements allow a system to predict system malfunctions/failures even before they occur. The current flow measurements also help the robotic modules to efficiently utilize the available energy distributed in the organism. An important ingredient in the electro-mechanical design of any autonomous system is its ability to withstand system failures. From dynamic power sharing perspective, the system design must have an innate fault tolerance, e.g., current limiters, fuses, switches, etc., to withstand the effects of endogenous and exogenous failures. The control over the current flow path in a modular robotic organism, at one end provides adaptive fault tolerance to an individual and collective systems, on the other hand, allows the robotic modules to efficiently utilize the distributed energy reserves in the organism, for instance, by forming sub-power buses. The chapter then introduced the platform design of state-of-the-art REPLICATOR re-configurable robotic modules. The electronic design of the REPLICATOR robotic modules, more explicitly, the power management system with fault tolerant energy sharing, has been developed and used for experiments in this research work. Later, it reviewed some of the bio-inspired techniques that have been applied in other robotic applications to gain energetic autonomy. The presented bio-inspired techniques include artificial foraging, trophallaxis, homeostasis and fault tolerance. At the end, some of the simulation tools used during the work were briefly described.

Chapter 3
Concept

Abstract This chapter presents the core concept that is conceived to address the issues related to self-sufficiency of a modular robotic organism. In the beginning it presents the open issues at a modular robotic organism from the perspective of self-sufficiency. To model the self-sufficient behavior of a modular robotic organism, firstly, the energetic modes of a self-sufficient robotic module are defined using a finite state machine. On the basis of energetics of a robotic module, the behavior model of a self-sufficient robotic module in a robot swarm is then defined. By combining the individual behavior of self-sufficient robotic modules it then defines the self-sufficient behavior of a modular robotic organism. The chapter then presents the concept of dynamic power management in a modular robotic organism. In this regard, getting inspiration from the homeostasis phenomenon found in biological systems, it introduces a novel concept of artificial energy homeostasis to achieve self-sufficiency at a modular robotic organism level.

3.1 Open Issues at a Modular Robotic Organism Level

As stated in problem description, from the perspective of a modular robotic organism, *self-sufficiency* can be defined as the collective ability of the robotic modules to find fuel and refuel, autonomously. This is only possible as long as the robotic modules retain enough energy to perform their role or responsibility in the organism. Here, from the perspective of self-sufficiency, attention is drawn to those critical system parameters, fixed or adaptable, that can influence the collaborative operations of the robotic modules in a modular robotic organism and are therefore required to be addressed. Following are the system states that can emerge in different energetic situations in a modular robotic organism;

- **Non-uniform energetic status of the robotic modules**
  This refers to a situation in which the energetic status of the robotic modules varies greatly in an organism. Such a situation normally arises under the following circumstances: for instance, just after an organism formation when the energetic status of the robotic modules was not taken into account prior to their docking, secondly, because of an uneven mechanical load distribution. That is, during an

R.H. Qadir, *Self-Sufficiency of an Autonomous Reconfigurable Modular Robotic Organism*, Adaptation, Learning, and Optimization 17,
DOI 10.1007/978-3-319-10289-4_3

organism locomotion, the mechanical load every robotic module experiences is dependent on its position in the organism's skeleton. In such a situation without power sharing, an organism may easily end up in a state in which although it carries enough energy, i.e., distributed among the robotic modules, but is no longer able to continue its operations.

- **Current flow through the organism's power bus**
  This is an important parameter that can be used as a measure to ensure the steady state operations within a modular robotic organism. The current flow through the organism's power bus depends on various parameters, e.g., the number of robotic modules in an organism, the morphology of the organism, the energetic status of the robotic modules, the power sharing/distribution mechanism, the topology of the organism's power bus, the collective task, etc. Therefore, it is essential to learn and specify the range of current flow values, an organism's power bus can have in different segments under different conditions.
- **Collective behavior of robotic modules in the presence of fault**
  This property refers to a situation in which a fault or a component failure at a robotic module causes deviations in the operations of multiple robotic modules in the organism. For example, a short circuit in a robot's electronics may also cause damages to the docked neighbor robotic modules. To deal with such situations, the robotic modules require innate and adaptive fault tolerance to instantly react and anticipate a possible fault condition, beforehand.
- **Fault prediction and identification**
  These are the two adaptive system features of an autonomous robotic system that essentially increase robustness in different operating conditions. The fault prediction allows a system to predict a possible fault situation, e.g., from the sensor measurements. In addition, fault identification may help in determining an appropriate response of the affected system.
- **Morphology**
  Considering the composition of a modular robotic organism, it is the morphology of the organism that directly and indirectly influences the sustainability and capabilities of the docked robotic modules. For instance, in a heterogeneous modular robotic organism, the morphology determines how the capabilities of individual robotic modules can be utilized in a cost effective and efficient manner. An unstable morphology not only limits the dynamic collaboration, but also increases the chances of potential failures among the robotic modules. For instance, a mechanically unstable robotic organism may require more resources than a stable one, e.g., power consumption, collective force during locomotion, computation resources, etc., to accomplish simple tasks in the arena. Therefore, during an organism formation, prior to the docking of a robotic module at a particular position in the organism, it is important to consider not only its energetic status but also the physical, sensing and computation abilities.
- **Cost of re-organizing the robotic modules in an organism**
  This parameter becomes important, when an organism has to adapt a different morphology either to overcome the external, i.e., environmental dependent constraints, or to get rid of a faulty robotic module. In both situations, it is important to

consider the re-organization cost in terms of net energy consumption for adapting a new morphology, that involves un-docking and re-docking of the robotic modules.

- **Distributed and centralized control mechanism**
  It is quite difficult to trade-off between centralized and decentralized control mechanisms. A common problem observed with centralized control mechanisms in other distributed systems is the "single point of failure" (SPOF). A SPOF is a part of the system, that in a case of failure can stop the entire system from functioning. In a modular robotic organism, it becomes difficult to establish a centralized control mechanism not only because of the single point of failure problem but in effect it may overload the designated robotic module(s). Alternatively, in a distributive control mechanism, the robotic modules in an organism will be required to periodically share their status information, e.g., energetic status, sensor measurements, etc., simply to synchronize their collective actions. This periodic information sharing may in result demand more computational power or resources at each robotic module in the organism.

To address the above mentioned issues that are mainly linked with the energetics of the robotic modules, a behavioral modeling approach is used to model the self-sufficient behavior of a modular robotic organism. It is chosen to focus on the challenges involved in the behavioral states that it may has to adapt to accomplish the self-sufficiency task.

3.2 Behavior Modeling

Behavioral modeling is commonly employed in many systems for early system design and analysis from multiple view points. Before modeling the behavior of an energetically autonomous modular robotic organism, it is important to clearly define the energetic states of a single robotic module with respect to its on-board energy reserves. An energetically autonomous mobile robot that owns the ability to replenish energy from the environment in a robot swarm undergoes different energetic states: *self-contained, sustenance, dozing*, and *dead*. Figure 3.1 shows the proposed finite state machine capturing the energetic states of a self-sufficient robotic module.

Self-contained Considering the energetic status of a robotic module, it is its healthiest state. A robotic module in this state has enough energy to perform the assigned tasks and can search, locate and dock to an available recharge station in the arena.

Sustenance It is the operating state in which the highest priority task for a robot is to "search, locate and dock" to a nearest available recharge station in the arena in order to sustain its energetic autonomy. A robot enters the *sustenance* state when its on-board energy reserve becomes insufficient to carry out the assigned tasks in the arena. Upon a successful recharge cycle, a robot can resume its operations left in self-contained state.

Dozing A robotic module loses energetic autonomy when it cannot successfully dock to a recharge station before its on-board energy reserves fall below a

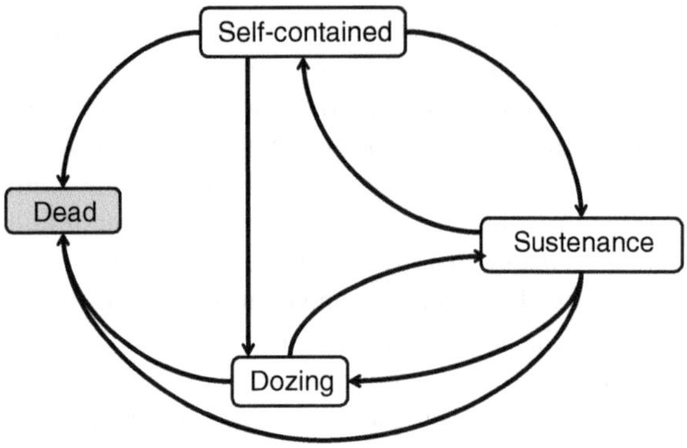

Fig. 3.1 Finite state machine of a self-sufficient robotic module with respect to its energetic states

pre-defined threshold limit. A *dozing* robotic module conserves its remaining battery charge by turning off its power hungry tasks or components, e.g., image processing, actuators, sensors, etc. A robotic module, to regain its lost energetic autonomy, in this state tries to seek help of other robotic modules in the arena, for instance, by periodically broadcasting distress messages or signals to the swarm.

Dead This state represents a condition in which a robotic module either due to insufficient battery charge or a fault is no longer able to continue its operation. That is, due to insufficient battery charge, the essential system components, such as controllers, sensors, etc., cannot be powered. In addition, a fault situation either due to a sudden breakdown, such as, a short circuit, or a faulty component, terminates the operations of an autonomous robot. A robot can only be revived from the dead state by human intervention.

3.2.1 Behavior Modeling of the Single Robotic Module

In the context of swarm robotics, the microscopic modeling approach, i.e., behavior modeling of the single robotic modules in a robot swarm, has been applied in a variety of applications, like Ijspeert et al. (2001), Martinoli and Easton (2003), Martinoli et al. (2004) used a probabilistic microscopic modeling approach to model the stick pulling problem in a robotic swarm, Lui (2008) used it to model the collective foraging task in a robotic swarm, and so forth. The approach captures the dynamics of a robotic module during its interaction with its swarm-mates and with the objects in the environment, but the details about its sub-components, e.g., sensors, actuators, body pose, are abstracted.

From the perspective of self-sufficiency, the applicability of *basic cycle of work* defined by McFarland and Spier (1997), is mainly dependent on two factors: the environment in which the robots are deployed or operating, and the degree or level of autonomy a robot owns. The *external factors* that are mostly beyond an individual's control include the terrain, number of available energy resources, number of robotic modules in a swarm, obstacles, etc. The *degree of autonomy* is the ability of an autonomous system to dynamically control, learn and regulate its various system parameters that influence its behavior in response to internal and external stimuli. To address these issues in the design of a robotic platform such factors are classified into two categories: *internal*—system dependent—and *external*—environment dependent.

The *internal factors* that are critical in gaining the energetic autonomy to an autonomous individual, solely depend on its electro-mechanical and software architecture. It includes, firstly, the "control over its locomotion". That is, how much power the actuators require to change its position? More explicitly, how much power a robot requires to overcome an obstacle or reach a recharge station in the arena? It is critical since the rest of the robotic module's functionalities are directly or indirectly dependent on its movement pattern in the arena. The second internal factor, is an individual's "sensing ability". The *sensing ability* of an individual is dependent on different sensor tools, e.g., microphone, vision system, laser, infra-red sensors, etc. Using these sensor measurements an autonomous individual can observe the changes continuously happening in its surroundings to tune or adapt its behavior accordingly. The third internal factor is the "self-regulating"ability. That is, from the perspective of self-sufficiency, the execution of tasks on the basis of on board energy reserve. For instance, if a recharge station became occupied while approaching it with a low energy reserve, the forager robot may choose to stop and wait for its turn instead of locating other recharge stations in the arena. And lastly, the ability of a system to withstand abrupt system malfunctions, such as a short circuit, a component failure, erroneous measurements, etc. With innate fault tolerance, embedded at the physical layer, a robotic module may save itself from unpredictable breakdowns or potential malfunctioning of the system components and may not effect other robotic modules when physically docked with them.

Considering the electromechanical design of a re-configurable robotic module in general, and a REPLICATOR module in particular, and its operational abilities in a robot swarm, a generic microscopic model was devised that captures the behavioral states it may adapt during its interaction with the swarm-mates and surrounding objects. Here, it is important to emphasize that the devised microscopic behavioral model serves as the basis for macroscopic behavioral modeling of a modular robotic organism. Figure 3.2 shows the finite state machine of an energetically autonomous robotic module. Each state models the behavior of a robotic module in a particular sequence. The label on each transition arrow between the states represents the conditions under which an individual robotic module changes its behavior. The microscopic model captures the behavior of an energetically autonomous robotic module using the following states: *nesting, searching, avoiding, approach and align, docked to a recharge station, docked to a robot, separation/un-docking, fainting, hibernation, broken/faulty*, and finally the *dead* state.

Nesting The *nesting* behavior of a robotic module represents it's healthiest energetic status in its life cycle. A *nesting* robot carries enough energy to explore the environment, perform the assigned task and if required can donate its

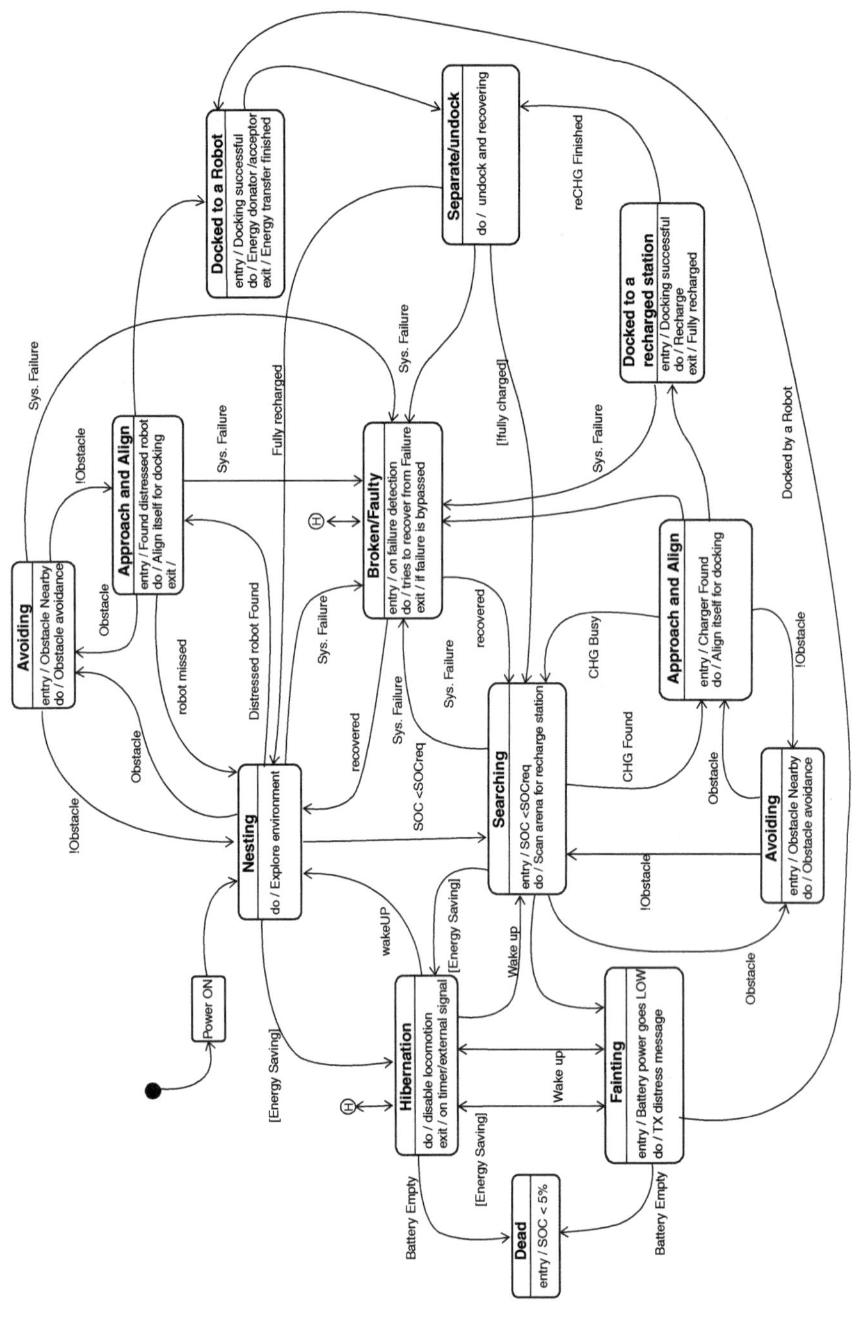

Fig. 3.2 Proposed FSM of an energetically autonomous robotic module

battery charge to a distressed robot in the arena. A *nesting robot* depending upon internal or external stimuli may adapt the following behavioral states: *avoidance, approach and align to a fellow robot, broken/faulty, searching a recharge station* or the *hibernation* state.

Searching This behavioral state is vital to a robotic module that has to keep its energetic autonomy. A robotic module actively starts searching an energy resource when its on-board energy reserve falls below a predefined or a learned "threshold limit". The threshold limit defines the amount of energy required to locate and successfully dock to an available recharge station in the arena. During the search for a recharge station a robot may adapt the following behavioral states: *broken/faulty, approach and align to a recharge station, avoidance, fainting*, or the *hibernation* state.

Avoidance An obstacle can either be a robot, a recharge station, an object or a boundary wall in the arena within the detection range of a robotic module, which forces it to enter the "avoidance" state for the period of time the obstacle remains in its detection range.

Approach and align This behavioral state is usually adapted by a robot either during *nesting*, when it finds a distressed robot in its vicinity or during *searching*, on locating a recharge station in the arena. In both scenarios, the particular robotic module moves towards and aligns itself for a successful docking.

Docked to a recharge station This behavioral state is attained when a robotic module has successfully docked to an available recharge station in the arena and draws current from the station to recharge its on-board battery.

Docked to a robot This behavioral state is achieved by robotic modules when an energetically healthy robot previously in the *nesting* state successfully docks to a robot broadcasting distress messages in the arena. To mimic the trophallaxis behavior of social insects, the energetically healthy robotic module donates a certain amount of its battery charge to the energetically weaker robotic module by recharging its on-board battery pack.

Separate/un-dock In this behavioral state, a robotic module already docked to either a recharge station or one of its teammates mechanically un-docks. For this purpose, firstly, it opens the current flow path on the particular docking interface and then issues an un-dock command to its docking drive controller.

Hibernation This behavioral state is added in the model solely for the purpose of energy conservation. An active robotic module may adapt this behavior either when the conservation of energy is desired, or it is idle for a certain period of time, i.e., has no assigned task to fulfill.

Fainting It is an indication that either there are fewer energy resources compared to the foraging robotic modules in the environment, or a robot fails to dock to an available recharge station in the arena. A robot enters the fainting state when its on-board energy reserve becomes insufficient to continue its locomotion in the environment. A faint robotic module therefore broadcasts "distress" messages in the swarm to attract "healthy" robotic modules for the purpose of regaining energy.

Broken/faulty Faults and failures can never be ignored in the design of any artificial system. This behavioral state is an indication that a fault or a failure either in the mechanics or in the electronics or in both has occurred. Depending on the occurred fault a robotic module tries to bypass it and if in case of success, it regains its previous behavioral state. Otherwise, it requires external assistance to revive its operations.

Dead This behavioral state indicates the end of an active life cycle of an autonomous robotic module. It appears when either a robotic module has not enough energy left to even remain in fainting or hibernation state or because of a sudden breakdown, the system becomes unresponsive. A dead robot requires external assistance for its recovery.

The proposed behavioral model of the single robotic module was used to conduct a case study that explored the foraging behavior of a robotic swarm in an arena with limited energy resources. The model in the simulation experiments tried to obtain the ratio between the number of recharge stations to the number of robotic modules that can reduce the number of dead robots in different swarm sizes (Humza and Scholz 2011).

3.2.2 Behavior Modeling of a Modular Robotic Organism

In the context of swarm robotics the behavior modeling approach focusing on capturing the behavioral aspects of the whole system, such as a robot swarm, is known as *macroscopic modeling* (Bayindir and Şahin 2007), briefly described in Sect. 2.2. Concerning the self-sufficient behavior of a modular robotic organism, before modeling the collective behavior of the robotic modules, it is important to identify the operating modes of a robotic module in an organism. From the perspective of dynamic power management, based on the energetic states, defined above, a robotic module in an organism can behave in either of the following operating modes: *donor, dependent*, or *neutral*.

- **Donor**
  The term *donor* or *energy donor* is designated to those robotic modules in an organism that at discrete time instant "*k*" are energetically healthy, for example, having $SOC_{curr}(k) > 50\,\% \cdot SOC_{max}$, and are able to donate their on-board battery charge to the energetically weaker robotic modules. Here, the value "50" is used as an exemplary threshold.
- **Dependent**
  The *dependent robotic modules* are those that either do not have enough energy left to power up their on-board system components or require more power to continue their assigned task in the organism. For instance, in an organism the robotic modules that are responsible for actuation in 2D and 3D typically require more power than the nominal for their collective locomotion.

- **Neutral**
  The term *neutral* is designated to those robotic modules that neither share nor take energy from other robotic modules in the organism. They are operating either in *partially faulty* or *fully broken* state.
  - **Partially Faulty**
    It is the state in which a fault in a system component causes deviations in the operations of a robotic module. The word *partially* here represents the system state of a robotic module that despite the faulty system components can still fulfill the assigned task in the organism. For instance, a failure in one of the vision systems can be by-passed to restore the steady state operations of a robotic module.
  - **Fully Broken**
    In this state, a component failure neither allows the affected robotic module nor the docked robotic modules to continue their operations in an organism. For example, a short circuit or a critical failure in the system electronics of a robotic module can also affect the docked robotic modules in the organism. In such a scenario, it becomes essential for the rest of the robotic modules in the organism to get rid of such a faulty robotic module either through physical re-organization of their structure or if possible by some other intelligent means, at the application software layer.

The devised behavioral model captures the behavior of a modular robotic organism from its energetic autonomy and fault tolerance perspective. Figure 3.3 shows the proposed behavioral model of a modular robotic organism. The states represent the collective behavior of multiple robotic modules in an organism that emerges from their local interaction and collaboration. In the similar manner, to maintain energetic autonomy a modular robotic organism in its life cycle adapts the following behaviors in different conditions: *nesting, searching, approach and align, avoiding, docked to a recharge station, separate/un-dock, reorganize, split, partial breakdown, full breakdown, hibernation*, and *dead.*

Nesting A modular robotic organism in this behavioral state carries enough on-board charge to perform the assigned tasks in the arena.

Searching In this behavioral state a modular robotic organism tries to maintain/keep its energetic autonomy that is based on the energetic status of the individual robotic modules. During *searching* the robotic modules of an organism collectively try to search and locate the available recharge stations in the arena. For this purpose, it may turn off redundant system components among the robotic modules for the purpose of energy conservation.

Approach and align Upon locating an available recharge station in the arena, the robotic modules in an organism collectively align themselves such that the robotic module facing the recharge station can easily dock to it.

Docked to a recharge station After a successful docking to an available recharge station in the arena, an organism depending on the number of robotic modules

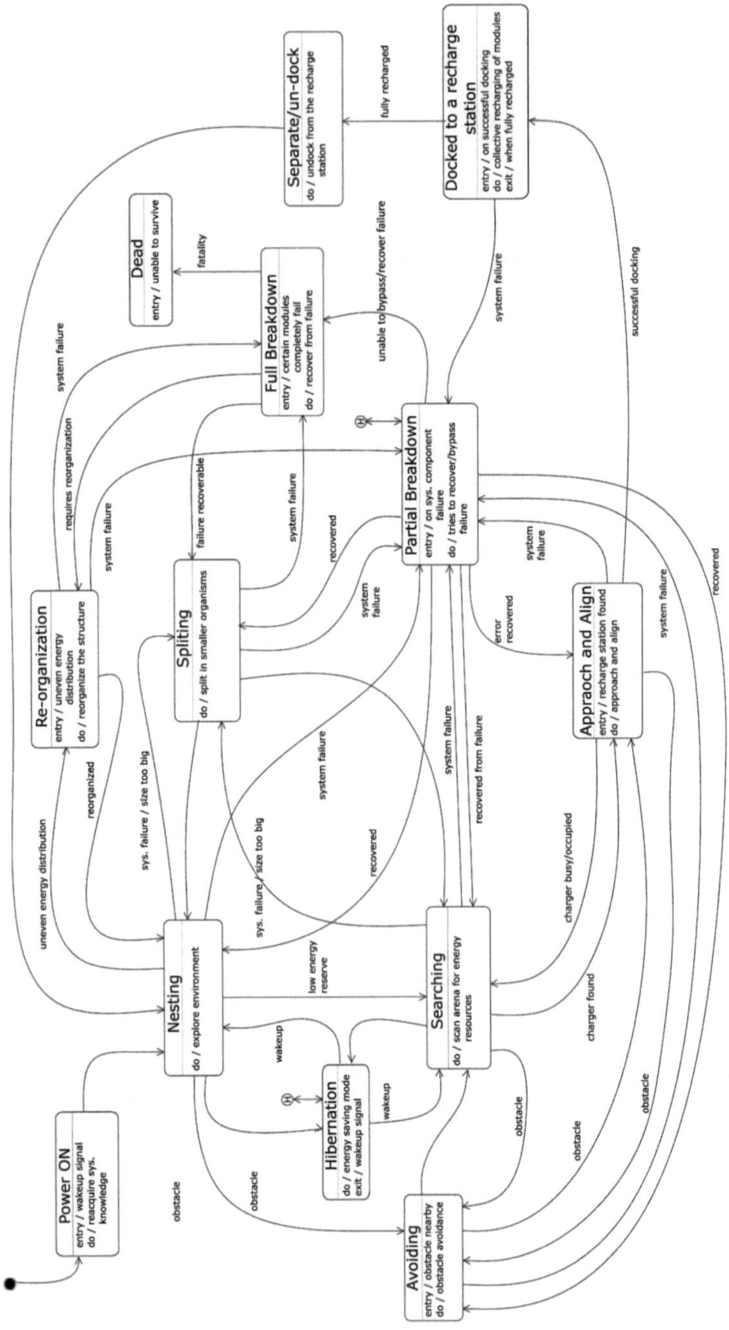

Fig. 3.3 Behavioral model of an autonomous modular robotic organism

and the organism's power bus capacity, collectively recharges its robotic modules from the recharge station.

Separate/un-dock This behavior results when an organism after successfully recharging its robotic modules un-docks/detaches itself from the recharge station to resume its operations in the arena.

Avoiding This behavioral state is taken upon the detection of any hindrance that appears in the detection range of an organism.

Re-organization An organism opts this behavioral state usually when a re-organization/re-structuring of the robotic modules in its skeleton is desired. An organism may re-organize the position of multiple robotic modules in its structure mainly to replace the faulty robotic modules or to meet the internal, e.g., energy distribution, sensing ability, etc., and external, e.g., obstacles, task, etc., requirements without adapting a new morphology.

Splitting An organism chooses this behavioral state when its size becomes too big and the available energy is not sufficient to continue the collaborative operations or to accommodate some environmental constraints. To accommodate such conditions an organism splits itself into smaller organisms to adapt a new morphology.

Partial breakdown The collective locomotion of robotic modules in an organism may in some scenarios increase the probability that different types of mechanical or electrical faults can occur. In response to a fault detection that can be fatal, the robotic modules in the organism may stop their operations and take this behavioral state.

The robotic modules in this state using their innate and adaptive fault tolerance features try to bypass the faults or faulty components in the system to restore their steady state operations. For example, from the perspective of power sharing, the robotic modules that are not directly affected by the occurred fault can form sub-power buses to resume power sharing in the organism.

Full breakdown A full system breakdown is an indication of a severe electromechanical failure at one or more than one robotic modules in the organism that cannot be bypassed. In such a situation, the robotic modules in the organism first tries to isolate the faulty/broken robotic module(s) by performing a series of predefined tests. After the detection and identification of an occurred fault, the organism reorganizes its morphology by discarding the faulty robotic module(s) or by splitting itself into multiple smaller organisms. In a more severe situation, a system failure may lead to a full breakdown.

A direct consequence of a full breakdown situation appears in the form of reorganization or splitting of an organism whereas in a partial breakdown situation it is not mandatory.

Hibernation An organism for the purpose of energy conservation undergoes hibernation for a certain amount of time. The particular behavioral state can be adapted either from "nesting" or "searching", by turning off its locomotion and unnecessary electronic components.

Dead This behavioral state represents a condition in which a modular robotic organism is no longer able to maintain or sustain its operations. From the perspective of energetic autonomy, this state appears when an organism either runs out of energy or an abrupt system failure paralyzes it. A dead organism therefore requires external assistance for its revival.

3.2.3 Formulation of the Organism's Behavioral Model

A mathematical model of a modular robotic organism involve multiple interdependent parameters of the robotic modules linked with each macroscopic behavioral state. From the perspective of self-sufficiency, the parameters include a robotic module's energetic status, operating state, position in the organism structure and its role in the organism's locomotion. For instance, in the case of a fault situation, the severity of an occurred fault on the steady state operations of a modular robotic organism can be determined by the position of the faulty robotic module in the organism structure, fault type and the role of the particular robotic module in the organism.

For a simple mathematical model that formulates the state transitions of the macroscopic model, consider a modular robotic organism which comprises N_{org} number of robotic modules, *nesting* in the arena. From the power sharing perspective, a robotic module in an organism is operating in either of the three modes: *donor*, *acceptor/dependent*, or *neutral*. Consider, at time instant "k", the variables $N_{dn}(k)$, $N_{ap}(k)$, and $N_{nu}(k)$ represent the number of *energy donor, energy acceptor* and *neutral* robotic modules, respectively, so that, the total number of robotic modules in a modular robotic organism is;

$$N_{org} = N_{dn}(k) + N_{ap}(k) + N_{nu}(k). \tag{3.1}$$

The variable $N_{nu}(k)$ represents the number of robotic modules that neither share their on-board energy reserve nor consume from other robotic modules, i.e.,

$$N_{nu}(k) = N_{pb}(k) + N_{fb}(k), \tag{3.2}$$

the sum of *partially* and *fully broken* robotic modules, respectively.

3.2.3.1 Transition from Nesting to Searching

Considering the energy reserves of a modular robotic organism, a transition from *nesting* to *searching* occurs when the on-board energy reserves become insufficient to further continue operations in *nesting* state. Similarly, as in the case of a self-sufficient robotic module, assume C_{res} is the amount of charge an organism should reserve to search and successfully dock to a recharge station in the arena. Then theoretically, a robotic module in the organism at time instant "k" must have at least,

$$C_j(k) = \frac{C_{res}}{N_{org}}, \tag{3.3}$$

amount of charge in average. Where, $j = 1, ..., N_{org}$. As described earlier, the robotic modules in an organism that can collaborate actively with respect to the energy sharing are the only ones operating without failures, i.e.,

$$N_{active}(k) = N_{org} - N_{nu}(k), \tag{3.4}$$

while, the rest of the robotic modules, i.e., $N_{pb}(k)$, suffering of faults are acting passively in terms of power sharing. Therefore, their on-board energy reserve cannot be included in the overall organism's charge calculation. With this assumption, Eq. (3.3) becomes,

$$C_j^{'}(k) = \frac{C_{res}}{N_{active}(k)}, \tag{3.5}$$

the amount of charge an *active* robotic module must possess in average to remain collectively in *nesting* state. The net organism's charge at time instant "k" can now be obtained as

$$C_{org}(k) = \sum_{j=1}^{N_{org}} C_j^{'}(k), \tag{3.6}$$

with the condition that jth robotic module is either an "energy donor" or "acceptor".

Using the above calculations, an organism from *nesting* opts *searching* behavior when its on-board charge reaches the threshold reserved for the *searching* task, i.e.,

$$C_{org}(k) < C_{res}.$$

While searching an energy resource in the arena an organism may split itself into smaller organisms, when it becomes difficult either to sustain the structure or to accommodate some environment driven constraints.

3.2.3.2 Transition from Nesting to Reorganization

In the case of an uneven energy distribution or faulty robotic modules, an organism can reshuffle the position of multiple robotic modules in its skeleton mainly to establish an even energy sharing between them. An uneven energy distribution situation may arise after an organism formation, for instance, when the robotic modules are docked to each other irrespective of their individual state of charge.

To model an uneven energy distribution scenario, consider a modular robotic organism that is logically divided into two halves. The collective charge of robotic modules in the two logical halves at time instant "k" can be obtained as,

$$C_{sb\_org1}(k) = \sum_{l=1}^{||N_{org}/2||} C_l(k), \tag{3.7}$$

$$C_{sb\_org2}(k) = \sum_{m=||N_{org}/2||+1}^{N_{org}} C_m(k), \tag{3.8}$$

where, variables $C_{sb\_org1}(k)$ and $C_{sb\_org2}(k)$ represent the on-board charge of the two logical halves of an organism, respectively. An organism may choose to reshuffle the robotic modules if the difference between the collective charge of robotic modules in the two logical halves varies greatly or is not within the threshold limits.

From the perspective of self-sufficiency, the proposed organism's behavioral model includes, to highlight, the behavioral states in which an organism will be required to adapt different modes of power sharing between the robotic modules. For instance, a *nesting* organism may freely use the available energetic resources to utilize all on-board tools to explore the environment or to accomplish different tasks in the arena. On the other hand, the same organism in *searching* or *partial breakdown* states may not adapt the similar power sharing topology among the robotic modules due to different reasons. To address the issues involved in power sharing between the robotic modules of an organism the following section presents the conceived aspects of dynamic power management in a modular robotic organism.

3.3 Power Management in a Modular Robotic Organism

In a standalone robotic system, dynamic power management or simply power management is a design methodology that provides control over the system's electronic components to reduce the overall power consumption without compromising the system performance and functionality (Benini and Micheli 1998; Lorch and Smith 1998). At an organism level, the power management can be realized through reactive and predictive response or capabilities of the docked robotic modules.

- **Reactive power management**
  The term *reactive* refers to a system's ability to produce an instant reaction, for instance, against abrupt system perturbations. The *reactive power management* features are typically embedded in the system design and therefore cannot be altered on the fly. A hardware current limiter in the electronic design is one such example that limits the current flow to a hard coded value (set point). In the case of a deviation from the system steady state, e.g., a short circuit, the current limiter automatically limits the current flow by opening the current flow path. In the context of a modular robotic organism, one such application scenario can be during a fault situation at a robotic module. That is, an abrupt system response to a faulty situation from the robotic modules in an organism may save them from potential damages.
- **Predictive power management**
  Contrary to reactive power management, *predictive* power management is based on prior system knowledge, i.e., system power consumption, number of robotic

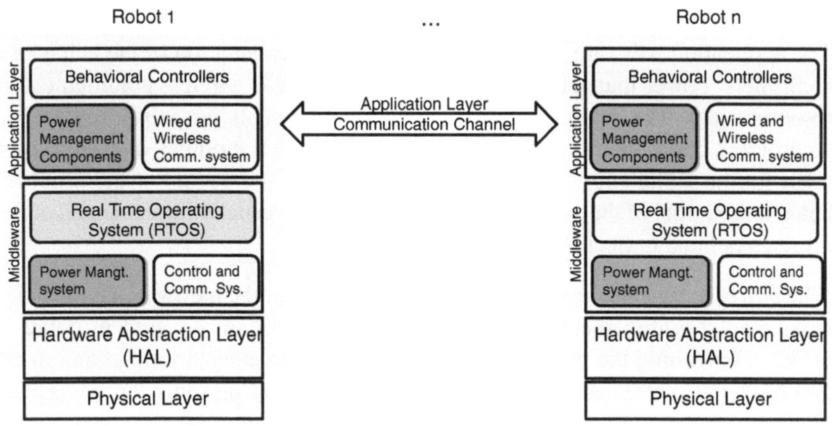

Fig. 3.4 Conceived layered architecture of a modular robotic organism

modules in an organism, energy distribution, etc. With the changing conditions, it tries to estimate or learn the new set points of its essential variables. The newly learned set points then allow an autonomous system to tune or choose its behavior to adapt the new operational conditions. An example scenario can be seen in case of redundant or unused system components. For instance, having an Ethernet controller in the active state without an open socket is a waste of energy. Therefore, for the sake of energy conservation the components should be turned ON only when their functionality is desired.

For the division of functionalities and software development, a layered architecture of a robotic module was proposed by the REPLICATOR consortium. Figure 3.4 shows the conceived layered architecture of a REPLICATOR organism. It is included to highlight the proposed power management components and a communication channel for sharing the local system information at the application layer. The layered architecture is divided into: *hardware abstraction layer*, *middleware* and *application layer*.

- **Hardware abstraction layer**
  This layer of software includes the low level device drivers and firmwares that are required to control the hardware components in the electronic design.
- **Middleware layer**
  The middleware layer is dedicated for the implementation of the real time operating system, basic power management features, communication and other control mechanisms. At this software layer, the power management system of a robotic module implements those essential features that are essential to keep an individual's energetic autonomy in its standalone mode, i.e., in a robot swarm. For this purpose, the details of the behavioral model of a single robotic module, i.e., the software related features as defined in Sect. 3.2.1, are implemented at the middleware.

- **Application layer**
  The application software layer is added for the implementation of the behavioral controllers, power management components, wired and wireless communication resources, etc. The implementation of such components is mainly required to control and coordinate the behavior of multiple robotic modules docked to each other in a modular robotic organism. In this regard, the details of the proposed behavioral model of the modular robotic organism are usually implemented at this layer, as defined in Sect. 3.2.2.

The *power management system* of a robotic module at the hardware layer, which was developed within the frame of this research work, are covered in detail in Sect. 4.2. Concerning the application software layer implementation, getting inspiration from the biological systems the following section presents the concept of homeostasis in the context of artificial systems.

3.3.1 Homeostasis in Artificial Systems

The details of homeostasis phenomenon has already been covered in Sect. 2.4.4. From a lower level, as identified by Owens et al. (2007), an artificial system that mimics the behavior of a biological homeostatic system must contain the following features:

- Prediction,
- Innate and adaptive response,
- Acclimatization.

The *predictive* ability of a homeostatic system allows it to estimate possible future states or an occurrence of an event through feed forward regulation. It enhances the system robustness by steering its behavior before the occurrence of an anomaly. *Innate response* of a self-regulating control system refers to those built-in system features that are embedded in the system design and therefore cannot be altered on the fly. Whereas, the *adaptive response* refers to the system behavior that is acquired or learned under different operating conditions. The innate response of the system design allows it to instantly react to the system anomalies. On the other hand, the adaptive response of an autonomous system allows it to adapt the changes that are necessary for its long-term survival. *Acclimatization* is a special form of adaptive response. It is a process in which an individual organism adjusts its internal behavior in response to gradual changes in its environment, such as, temperature, humidity, etc., in order to maintain performance and survive across a range of environmental conditions.

For the increased robustness and stability of the re-configurable robotic modules, the designed control system mimics the components of a biological homeostatic control system to maintain equilibrium in the presence of intra-cellular and intra-organism system perturbations. Figure 3.5 shows the block diagram representation of the application software layer control system of a robotic module. It includes a *control center, sensors* as receptors and *actuators* as the effectors. The *control center*

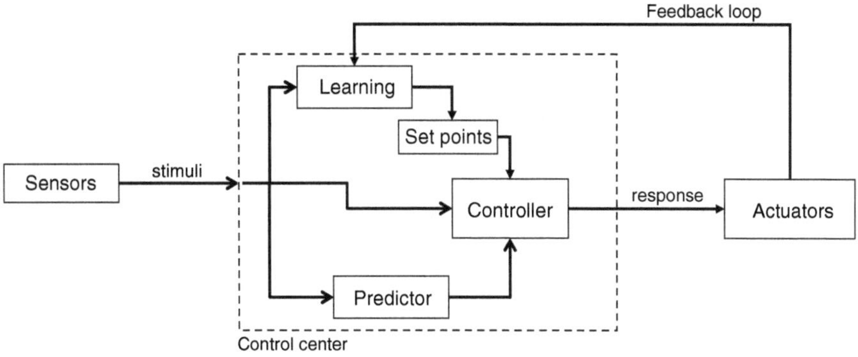

Fig. 3.5 Block diagram representation of the devised control system of a robotic module

further includes three distinct components: *learning*, *predictor* and the *controller*. The *learning* component samples the "stimuli" from the sensors and "feedback" from the actuators to learn the system *set points*. The *set points* define a range of steady state values, either fixed or dynamically learned, required to regulate the system behavior. The *predictor* component uses the sensor measurements to predict or estimate the system future state. The *controller* takes the sensor measurements and an estimate of the future state from the *predictor* component as input, to regulate the system behavior within the tolerance limits, defined by the set points.

The designed control system of a robotic module without the *predictor* component has been implemented in Stage simulator to explore the energetic behavior of the robotic modules in an arena with limited energy resources. The devised foraging behavioral model in that particular scenario tried to learn and adapt a predefined set point that forces a *nesting* robotic module in a robotic swarm to adapt *searching* behavior. The details of the implementation of the concept was published in Scholz (2011).

Considering the above mentioned open issues and the behavioral states of a self-sufficient modular robotic organism, getting inspiration from the homeostasis phenomenon in biological systems, a means of establishing an energetic equilibrium between the robotic modules of an organism is devised.

3.3.1.1 Artificial Energy Homeostasis

To prolong and strengthen the collaboration of multiple robotic modules without compromising the energetic autonomy of individual robotic modules, the devised control system focuses on establishing energy homeostasis in a modular robotic organism. The term *artificial energy homeostasis* in the context of a modular robotic organism is defined as

> A process that regulates the dynamic power flow in an artificial modular robotic organism by taking into account the energy distribution among the robotic modules, their role and position in the organism. Furthermore, to ensure the system equilibrium and robustness against potential faults and failures, it establishes adaptive fault tolerance mechanisms.

From the implementations perspective, the devised control system at each robotic module in a modular robotic organism regulates the *current flow* within the on-board system components and between the robotic modules, as the *essential variable*. The *set points*, a range of steady state current flow values, in a modular robotic organism generally vary under different load situations and therefore can be learned to accommodate the changes in the internal and external operating conditions. In case of a deviation from the set points, the control system, to restore the system steady state, initiates the fault detection and identification mechanisms for the isolation of an occurred fault.

The implementation of the two components of artificial energy homeostasis, power sharing and fault tolerance, at the application software layer of a robotic module are covered in detail in the following chapter.

3.4 Summary

This chapter presented the core concept of the work that was conceived to address the issues related to self-sufficiency of a modular robotic organism. In the beginning, the chapter described the open issues that are foreseen at a modular robotic organism level in different situations. To address these challenges at first the behavior model of a self-sufficient robotic module is defined. Based on the behavioral states of individual robotic modules in an organism, a self-sufficient behavior of a modular robotic organism is then defined. From the perspective of self-sufficiency, the proposed organism's behavioral model highlighted the behavioral states in which an organism will be required to adapt different modes of power sharing between the robotic modules. It then presented the concept of dynamic power management in a modular robotic organism. In this regard, a novel concept of artificial energy homeostasis to achieve self-sufficiency at a modular robotic organism level is introduced.

Chapter 4
Implementation

Abstract This chapter presents the implementation of the concept, both at the hardware and application software layer in a simulation framework. It begins with the hardware design considerations from the perspective of a robotic module's platform design. In this regard, the factors that influence the design of a dynamic power management system are explored in detail, that include it's electronic architecture, system power budget calculation, choice of system source voltage, and platform specific constraints. The chapter then presents the detail description of the proposed power management system with fault tolerant energy sharing and its design considerations. The proposed power management system has been developed and realized as a real microprocessor based hardware together with a dedicated firmware. From the perspective of application software layer implementation, firstly, it presents the devised simulation framework and its implementation details. It then presents the details of the proposed application software layer power management components of a robotic module required for dynamic power sharing in a modular robotic organism, and the application software layer fault tolerance at an organism level. At the end, a short summary concludes the presented work.

4.1 Hardware Design Considerations

To mimic even a simple swarm behavior of a social insect colony, observed in nature, the electro-mechanical system design of an artificial mobile robotic system requires to integrate various tools. The different tools, such as, sensors, actuators, vision system, etc., are required to address several challenges that emerge during an individual's interaction with its teammates and the environment. As mentioned in Sect. 3.2.1, the internal factors that are considered to be vital to achieve energetic autonomy, directly or indirectly effects the mechanical and electronic architecture of a robotic system. For example, the locomotion ability on a particular terrain is dependent on the type of actuators in the platform design.

In a battery powered mobile robotic system, among various, a significantly important parameter is the amount of energy reserve it can carry on-board, i.e. maximum charge capacity of the battery pack. In other words, the higher the battery charge

R.H. Qadir, *Self-Sufficiency of an Autonomous Reconfigurable Modular Robotic Organism*, Adaptation, Learning, and Optimization 17,
DOI 10.1007/978-3-319-10289-4_4

capacity the longer a robot can survive or continue its autonomous operations. But unlikely, in the system design there exists several other parameters that directly or indirectly have influence on the operational time of a mobile robot. Therefore, before a concrete system design, it is important to gain a clear and broader understanding of these system dependent parameters. Within this research work the platform design of REPLICATOR robotic modules having no power management system have been used as starting point for development of an own power management system with fault tolerant energy sharing. In this regard, firstly, the electronic architecture, common among all REPLICATOR robotic modules, is explored.

4.1.1 REPLICATOR Robot Electronic Architecture

Here, it is important to introduce the robot electronic architecture as in the following subsections the calculations made include the electronic components present in the REPLICATOR robot platform design. Figure 4.1 shows a block diagram of the electronic architecture that is more or less common on the three REPLICATOR robot platforms. To fully control and efficiently utilize the system capabilities the electronic architecture of the REPLICATOR robotic modules is divided into two control units, the "core" processor and the "peripheral" controllers. The *core processing unit* of each of the robot platform is a *CM-BF561* drop-in module from BLUETECHNIX equipped with a dual core *Blackfin BF561* μcontroller from ANALOG DEVICES. Via a Serial Peripheral Interface (SPI), this unit is attached to 4 peripheral μcontrollers (*MSP430F2618* from TEXAS INSTRUMENTS), which are responsible for sensor data acquisition, low level actuator control and processing in order to take off burden from the main processing unit. These peripheral controllers mainly serve as an interface between the application software routines running on the *Blackfin* processor and the system components. All μcontrollers and corresponding peripheral elements are placed on separate PCBs, installed on each side of a robotic module. Each of these PCBs have in common certain sensors and actuators, e.g. docking sensors and actuators, microphones, RGB-LEDs, etc., but may in addition take over specialized tasks, e.g. 2D or 3D locomotion, ZigBee radio interface, etc. For 3D actuation, up to two additional *LM3S8962 Cortex* μ controllers (LUMINARY MICRO) are integrated to permit high-performance brush-less DC-motor control. On each of the PCBs, a local I^2C bus is implemented for interfacing the local controller with the respective sensors. In addition, a global I^2C bus has been implemented to facilitate multi-master communications between the peripheral controllers.

The overall control mechanism of a REPLICATOR robotic module is divided into following blocks:

- **Distributed control**
  The peripheral controllers on each lateral side distributively control and manage the information, they collect from the sensors and system components present on the respective side.

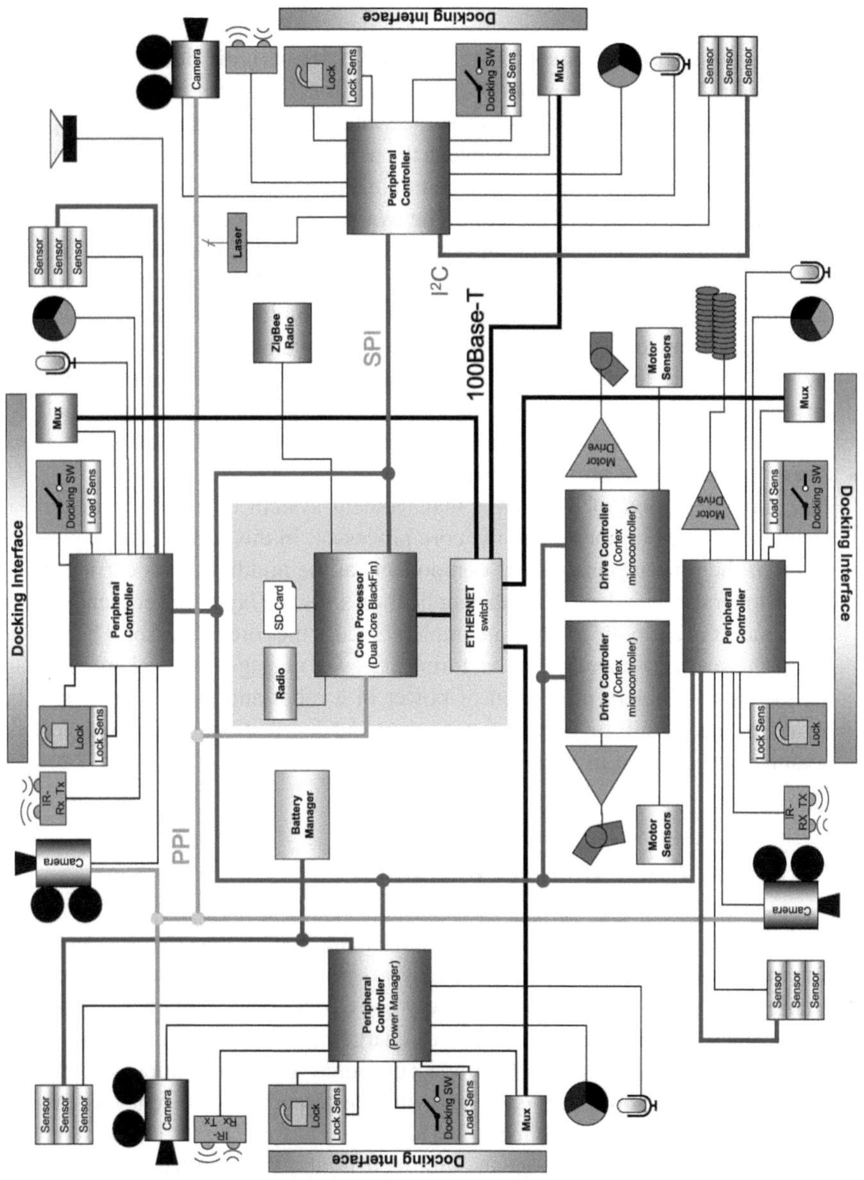

Fig. 4.1 Electronic architecture block diagram that is more or less common on the three REPLICATOR robotic platforms: Kabot, Scout robot and active wheel

- **Centralized control**
  The core processor gains the hardware control and information from the peripheral controllers to centrally monitor and control the behavior of a robot in "swarm" as well as in "organism mode".

- **Communication**
  The communication system of the REPLICATOR robots is divided into two blocks, the "intra-robot" and "inter-robot" communication. The *intra robot* communication involves communication between the system components and the peripheral controllers and communication between the peripheral controllers and core microprocessor. Whereas, the *inter-robot* communication uses two types of communication channels: wired and wireless. The *wired communication medium* involves 100BaseT Ethernet between the physically docked robotic modules. The Ethernet medium allows the robots to distributively process and share the heavy sensory information, e.g. vision, acoustic, etc., sampled simultaneously at various modules in the organism. The *wireless communication* enables the robotic modules in the swarm mode to share the necessary information about themselves and their surroundings with each other. For wireless communication the robotic modules are equipped with a $Zigbee^{TM}$ radio link and, additionally, infra-red transceivers.
- **Power management and distribution**
  The implementation of the power management system of a robotic module is divided over the peripheral and the core processor. In this regard, the implementation of the power management components at the middleware is carried out at the peripheral controllers. And, the application software layer power management components of a robotic module are implemented at the core processor. The robotic modules use reactive and proactive forms of power management and energy sharing to ensure efficient utilization of power in a robot and in a modular robotic organism. The details on power management and distribution in a modular robotic organism are covered in Sect. 4.4.

4.1.2 System Power Budget Calculation

For an autonomous battery powered mobile robotic system besides other essential requirements, a desiring element is the long term operational time with a single charge (assuming it is using a rechargeable battery pack). The long term operational time that very much depends on the battery pack charge capacity, is also influenced by its source voltage. Before discussing the impact of source voltage, it is important to know the component wise power consumption with their respective voltage levels, present in the system's electronic architecture. The *system power budget* calculation lists all the electronic system components with their minimum, nominal and maximum current ratings to estimate the overall system power consumption. Table 4.1 shows the system power budget of a REPLICATOR robotic module.

4.1.3 Choice of Source Voltage

The *system voltage* is the battery pack terminal voltage available to the system components. A direct impact of the system voltage on the overall system's energetic

Table 4.1 System power budget of a REPLICATOR robot

Components	Voltage (in V)			Current (in mA)		
	Min.	Nominal	Max.	Min.	Nominal	Peak
2D locomotion drives	6.0	12.0	24.0	200.0	240.0	748.0
2D drive controller	3.0	3.3	3.6	56.0	66.0	110.0
3D hinge drive	18.0	22.0	24.0	300.0	800.0	2000.0
3D drive controller	3.0	3.3	3.6	56.0	66.0	110.0
Drive retraction	6.0	12.0	24.0	41.0	195.0	788.0
4x Docking drives	6.0	12.0	24.0	41.0	195.0	788.0
Laser scanner	2.7	3.3	5.0	30.0	40.0	50.0
Radio (ZigBee) module	2.7	3.3	3.6	10.0	16.0	40.0
4x Docking sensors	3.0	3.3	3.6	40.0	40.0	200.0
4x RGB status LEDs	3.0	3.3	3.6	80.0	120.0	320.0
Localization unit	2.3	2.5	3.0	0.4	2.7	200.0
Core processor	3.0	3.3	3.6	61.0	103.0	475.0
4x Peripheral controllers	3.0	3.3	3.6	10.0	24.2	25.0
Ethernet module	3.0	3.3	3.6	600.0	757.0	975.0
4x Vision sensors	3.0	3.3	3.6	0.0	18.0	20.0
Power management module				10.0	15.0	20.0
Battery management module				0.8	1.0	2.0
Total current consumption				1546.2	2703.9	6871.0

efficiency can be measured at step-up and step-down voltage regulators. A step-up voltage regulator or boost converter is required if the source voltage is below the nominal voltage required to drive a component. Similarly, if the source voltage is high enough then a step-down voltage regulator or buck converter brings down the system voltage to the voltage level required. A direct impact of a low system voltage in a modular robotic organism appears in the form of high current flow during power sharing between the robotic modules that consequently increases the power losses.

In such circumstances, to select an appropriate system voltage level the current consumption of the system components listed in Table 4.1, are transformed with different source voltage levels. Table 4.2 provides the current consumption of the

Table 4.2 Transformed current consumption of individual components listed in Table 4.1 with three different on-board power sources

Components	Curr. transformed with 7.4 V			Curr. transformed with 11.1 V			Curr. transformed with 22.2 V		
	Min.	Typical	Peak	Min.	Typical	Peak	Min.	Typical	Peak
2-D locomotion drive	324.32	389.19	1212.97	216.22	259.46	808.65	108.11	129.73	404.32
2-D drive controller	24.97	29.43	49.05	16.65	19.62	32.70	8.32	9.81	16.35
3-D hinge drive	891.89	2378.38	5945.95	594.59	1585.59	3963.96	297.30	792.79	1981.98
3-D drive controller	24.97	29.43	49.05	16.65	19.62	32.70	8.32	9.81	16.35
Drive retraction	66.49	316.22	1277.84	44.32	210.81	851.89	22.16	105.41	425.95
4x docking drives	66.49	316.22	1277.84	44.32	210.81	851.89	22.16	105.41	425.95
Laser scanner	13.38	17.84	22.30	8.92	11.89	14.86	4.46	5.95	7.43
Radio (ZigBee) module	4.46	7.14	17.84	2.97	4.76	11.89	1.49	2.38	5.95
4x docking sensors	17.84	17.84	89.19	11.89	11.89	59.46	5.95	5.95	29.73
4x RGB status LEDs	35.68	53.51	142.70	23.78	35.68	95.14	11.89	17.84	47.57
Localization unit	0.13	0.90	67.57	0.08	0.60	45.05	0.04	0.30	22.52
Core processor	27.20	45.93	211.82	18.14	0.62	141.22	9.07	15.31	70.61
4x shadow controllers	4.46	10.79	11.15	2.97	7.19	7.43	1.49	3.60	3.72
Ethernet module	267.57	337.58	434.80	178.38	225.05	289.86	89.19	112.53	144.93
4x Vision sensors	0.00	8.03	8.92	0.00	5.35	5.95	0.00	2.68	2.97
Power management module	10.0	15.0	20.0	10.0	15.0	20.0	10.0	15.0	20.0
Battery management module	1.0	1.0	1.0	1.0	1.0	1.0	1.0	1.0	1.0
Total current consumption (mA)	1780.6	3974.2	10838.74	1190.7	2654.8	7232.50	600.0	1335.4	3626.25

system components transformed to three different source voltage levels. The chosen power sources are the different combinations of serially connected lithium polymer (LiPo) cells, e.g. a 2S1P configuration denotes a combination of 2 serially connected cells that provide a nominal voltage of 7.4 V. The first three columns, i.e. from two to four, in Table 4.2 provide the transformed current consumption of the individual components with a source voltage of 7.4 V. With a 7.4 V power source the typical current consumption of the system when all the components are in active state varies roughly between 4 A and 11 A, at most. The next three columns, i.e. from fifth to seventh, in the Table 4.2 show the measurements made with a source voltage of 11.1 V. With a 11.1 V power source the instantaneous system current consumption varies between 2.6 A and 7.2 A. And, lastly, the last three columns in Table 4.2 show the transformed current with a voltage source of 22.2 V. With a 22.2 V power source that results with a combination of 6 LiPo cells, the overall system current consumption varies between 1.3 A and 3.6 A. It is important to consider here, that these theoretical calculations do not include the efficiency of the system electronic components, i.e. the efficiency of step-up and step-down voltage regulators, and the current losses in the system.

The calculations obtained in Table 4.2 provide a rough estimate of instantaneous current consumption of a robotic module with three different on-board source voltage levels irrespective of the battery pack charge capacity. To obtain a more precise estimate of system's instantaneous current consumption an important parameter left over in the current calculations is the inclusion of duty cycle. A *duty cycle* is the percent of time a particular component remains in active state with respect to the operating mode of a robot, i.e. swarm or organism mode.

Table 4.3 provides the instantaneous current consumption of a REPLICATOR robotic module operating in swarm and organism mode again with three different on-board source voltage levels. The second column in Table 4.3 from left shows the duty cycle of components that are assumed in active state during the swarm mode of a robot. For instance, the duty cycle "0" for the 3D hinge drive in swarm mode shows it's permanent inactive state. Similarly, the third column shows the duty cycle of components during the organism mode of a robotic module. In swarm mode with a 7.4 V power source the instantaneous system current consumption is $\approx$ 700 mA and in organism mode it is raised to 2 A. In case of 11.1 V power supply the instantaneous system current consumption reduces to $\approx$ 500 mA and in organism mode it is decreased to 1.3 A from 2 A. Lastly, with a 22.2 V power source the system's instantaneous current consumption further reduces to $\approx$ 250 mA in swarm mode and in organism mode it ends up with approx. 650 mA.

These theoretical calculations provide an insight of the system electronic architecture in terms of its current consumption in swarm and organism mode at three different source voltage levels. But, in a battery operated *re-configurable modular robotic system* the system voltage selection criterion is not just limited to the calculations made in Tables 4.2 and 4.3, rather it counts on few other parameters that are dependent on the mechanical architecture of the robot platform.

Table 4.3 Current consumption of a REPLICATOR robotic module operating in swarm and organism mode with three different on-board power sources

Components	D. cycle	D. cycle	OC @ V_1	OC @ V_1	OC @ V_2	OC @ V_2	OC @ V_3	OC @ V_3
	Swarm	Organism	Swarm	Organism	Swarm	Organism	Swarm	Organism
2-D locomotion drive	0.9	0.0	350.27	0.0	233.51	0.0	116.76	0.0
2-D drive controller	0.9	0.0	26.49	0.00	17.66	0.0	8.83	0.0
3-D hinge drive	0.0	0.5	0.0	1189.19	0.0	792.79	0.0	39.40
3-D drive controller	0.0	0.5	0.0	14.72	0.0	9.81	0.0	4.91
Drive Retraction	0.5	0.5	158.11	158.11	105.41	105.41	52.70	52.70
4x docking motors	0.2	0.2	63.24	63.24	42.16	42.16	21.08	21.08
Laser scanner	0.3	0.3	5.35	5.35	3.57	3.57	1.78	1.78
Radio (ZigBee) module	0.8	0.5	5.71	3.57	3.81	2.38	1.90	1.19
4x Docking Sensors	1.0	1.0	17.84	17.84	11.89	11.89	5.95	5.95
4x RGB status LEDs	0.5	0.5	26.76	26.76	17.84	17.84	8.92	8.92
Localization unit	1.0	1.0	0.90	0.90	0.60	0.60	0.30	0.30
Core processor	1.0	1.0	45.93	45.93	30.62	30.62	15.31	15.31
4x shadow controllers	1.0	1.0	10.79	10.79	7.19	7.19	3.6	3.60
Ethernet module	0.0	1.0	0.0	337.58	0.0	225.05	0.0	112.53
4x vision sensors	0.5	0.5	0.0	0.0	0.0	0.0	0.0	0.0
Power management module	1.0	1.0	15.00	15.0	15.0	15.0	15.0	15.0
Battery management module	1.0	1.0	1.0	1.0	1.0	1.0	1.0	1.0
Total current consumption (mA)			726.38	1889.98	489.26	1264.31	252.13	639.66

Operational current, $V_1 = 7.4$ V, $V_2 = 11.1$ V, $V_3 = 22.2$ V

4.1.4 Platform Specific Constraints

In the electronic design of a re-configurable autonomous system, the critical system components that are affected by a robot's mechanical architecture include "docking connectors" and the "battery pack". The docking connectors at the docking interfaces are required for establishing a power bus between physically docked robotic modules. And, the selection of an appropriate battery pack that at one end depends on the choice of system voltage is also influenced with the available space on the robot platform.

4.1.4.1 Docking Connectors

The docking connectors (pins) are an influential component since their selection at one end is highly dependent on the mechanical construction of a robotic platform, on the other end, directly influences the electrical characteristics of the whole system, i.e. the amount of current that can flow between the docked robotic modules.

Concerning the hardware design of an autonomous robotic module, the parameter interlinked with the choice of docking connectors is the "system voltage". Here its importance is because of the components that are available on the market operating in the particular voltage range. Other considerations in regards to the system voltage level include, how much current

- will be required to flow through the organism's power bus?
- the docking pins can withstand?
- will be safe to flow through the organism's power bus, i.e. the upper limit that ensures the steady state operation of robotic modules within an organism?
- or power can a robot donate or share to other docked robots in an organism?

Theoretically, these parameters can be chosen arbitrarily. But, in a real system design they may limit certain essential features of an autonomous system. The answer to the first question can be obtained from the calculations made in Table 4.3, that provides a rough estimate of how much current can flow between the docked robotic modules during power sharing, when a robotic module is dependent on the other. The second question is solely a hardware driven factor that depends on the type of a docking pin. Its internal resistance and the power rating define the efficiency and current carrying capacity of the power bus, respectively. The third parameter depends on the power management system design and the number of robotic modules docked in an organism. And lastly, the fourth parameter directly depends on the battery discharge capacity and is indirectly coupled with the third question.

Considering the above mentioned platform specific constraints, the chosen docking interface pins that were found to be suitable for the REPLICATOR docking unit are shown in Fig. 4.2. Among the 9 docking pins, 5 pins (4 in diagonal and a central) are dedicated to the power bus and the rest of the 4 pins in the circle are used as Ethernet lines. The electrical specification of the chosen docking pins set an upper limit of 8 A of current through the organism's power bus. With the said limit on the inter-modular current flow, consider a scenario of collective recharging— recharging of

Fig. 4.2 Docking unit with power management system components on one side of Kabot

several robotic modules simultaneously in an organism —assuming each robot consumes 500 mA of charge current from the organism's power bus, then theoretically, no more than 16 REPLICATOR robotic modules can be recharged in parallel.

4.1.4.2 Battery Pack

The on-board space available for a battery is critical, since in case of a rechargeable Lithium-ion battery pack it directly determines its charge capacity typically measured in milli-Ampere hours (mAh) and its terminal voltage. The terminal voltage of a rechargeable lithium-ion battery pack depends upon the number of serially connected cells, and its charge capacity depends on a cell's size.

In rechargeable battery technology, lithium-ion cells are regarded as power efficient with respect to their high energy/charge density, low self-discharge rate, and high discharge current. Lithium-ion polymer or more commonly lithium polymer (Li-Poly) batteries are a subclass of Li-Ion battery with the added advantage of high energy density in a thin form factor. Therefore, those were chosen for the REPLICATOR robotic modules. Table 4.4 shows the electrical specifications of a typical Li-Poly cell. Where "C" corresponds to the maximum charge capacity of a Li-Poly cell.

Table 4.5 provides an estimate of operational time of a REPLICATOR robotic module with three different power sources each having a charge capacity of 1, 000 mAh and 75 % efficiency. From the calculations, a REPLICATOR robotic

Table 4.4 Electrical specification of a Li-Poly cell

Voltage (V)			Maximum charging current	Discharge current	Recharging cycles
Min.	Nominal	Max.			
2.8	3.7	4.2	$\leq$1 C	>15 C	>500

Table 4.5 Operational time of a REPLICATOR robot with three different power sources

Time/Mode	Oper. Time with 2S1P		Oper. Time with 3S1P		Oper. Time with 6S1P	
	Swarm	Organism	Swarm	Organism	Swarm	Organism
In minutes	61.95	21.59	91.97	32.28	178.48	63.87
In hours	1.03	0.35	1.53	0.53	2.97	1.06

module with a 2 cells Li-Poly battery pack (2S1P) can operate in swarm mode for about an hour and organism mode for ≈22 min. A REPLICATOR robotic module with a 3 cells Li-Poly battery pack (3S1P) can operate in swarm mode for ≈92 min, and in organism mode for ≈32 min. Lastly, a robotic module with a 6 cells Li-Poly battery pack (6S1P) can achieve a maximum operational time of ≈3 hours in swarm mode, and in organism mode for ≈1 hour.

Considering all the calculations made above, i.e. system power budget, choice of system voltage, system specific constraints, the estimated operational time, the available battery capacities, their weight and size of a robotic module, a battery pack of 6 Li-Poly cells was chosen. It provided the nominal voltage of 22.2 V with a maximum charge capacity of 1,050 mAh.

At the first step, after deciding the system voltage it was quite challenging to come up with a design of battery management module that can handle the operations of a six cells Li-Poly battery pack, e.g. charging and discharging, cell balancing, protection against over charge and discharge, estimating the battery pack state of charge, charge cycles, etc., and provide an interface to control and access these parameters at the application software layer of a robotic module. It became a challenging task because no such solution was available that can be used out of the box. To address the issues related to the battery management an AFE—BQ77pl900 —from Texas Instruments (TI), an only available solution, was used. And, to control and access the operations of the chosen AFE an ultra-low-power micro-controller—MSP430F2132 from TI— was selected, mainly because of its low power consumption and a small footprint. In the next step, a custom designed printed circuit board (PCB) was developed to put both the ICs in a desired configuration, as shown in Fig. 4.3. After finishing the hardware development, firmware was then written to control the operations of AFE in a host control mode and provide the desired information to the system's *power manager*.

Similarly, other challenges that were critical in the hardware development phase include the design of an appropriate battery charge mechanism that can support the concept of trophallaxis, the design of fault tolerant energy sharing and docking interfaces, putting all power management components on custom designed PCBs, programming of system *power manager*—peripheral controller—to control the functionalities of the power management components at the hardware layer and provide an access of low level information to the system's core processing unit.

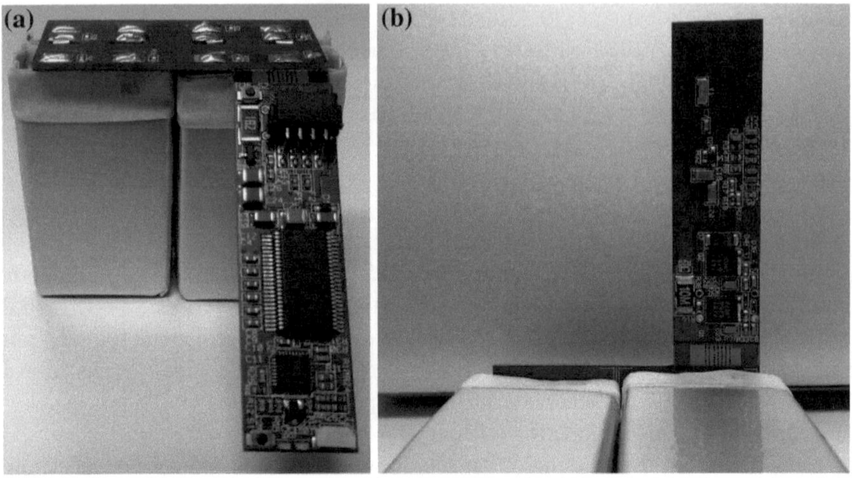

Fig. 4.3 Custom designed 6 Li-Poly cells battery pack electronics. **a** Battery pack electronics front side, **b** Battery pack electronics rear side

4.2 Proposed Power Management System—Hardware Design

It is a little difficult to provide an exact definition of a power management, since it's priorities or functions may slightly vary from application to application. In the context of swarm robotics, *power management* may refer to planning, generation, storage, distribution and dissipation of power in a system (Humza and Scholz 2010). It can be realized as a control system with an active feedback mechanism that monitors the environment (energy resources) and the system load (power consumption) to help adapt a system's behavior accordingly. The feedback mechanism allows an individual to maintain its operations within a steady state for a longer period of time while operating in a dynamic environment, e.g. limited number of energy resources, number of competitors, obstacles, etc.

The proposed power management system with fault tolerant energy sharing, used in REPLICATOR robotic modules, includes all the desired features that were found to be missing in the system design of state-of-the-art re-configurable robotic systems, as mentioned in Sect. 2.6. The proposed design not only includes the desired features in the hardware design, but also provides a higher layer access to the application software components to control and synchronize an individual's behavior with the docked robotic modules. Figure 4.4 shows the block diagram of the proposed power management system with fault tolerant energy sharing. It includes a battery management module, a battery re-charging unit, so called ideal diodes for uni-directional low loss current flow, a system power manager, an energy sharing module, a high power control switch for powering on-board peripherals, and four docking interfaces, one on each lateral side of the robot.

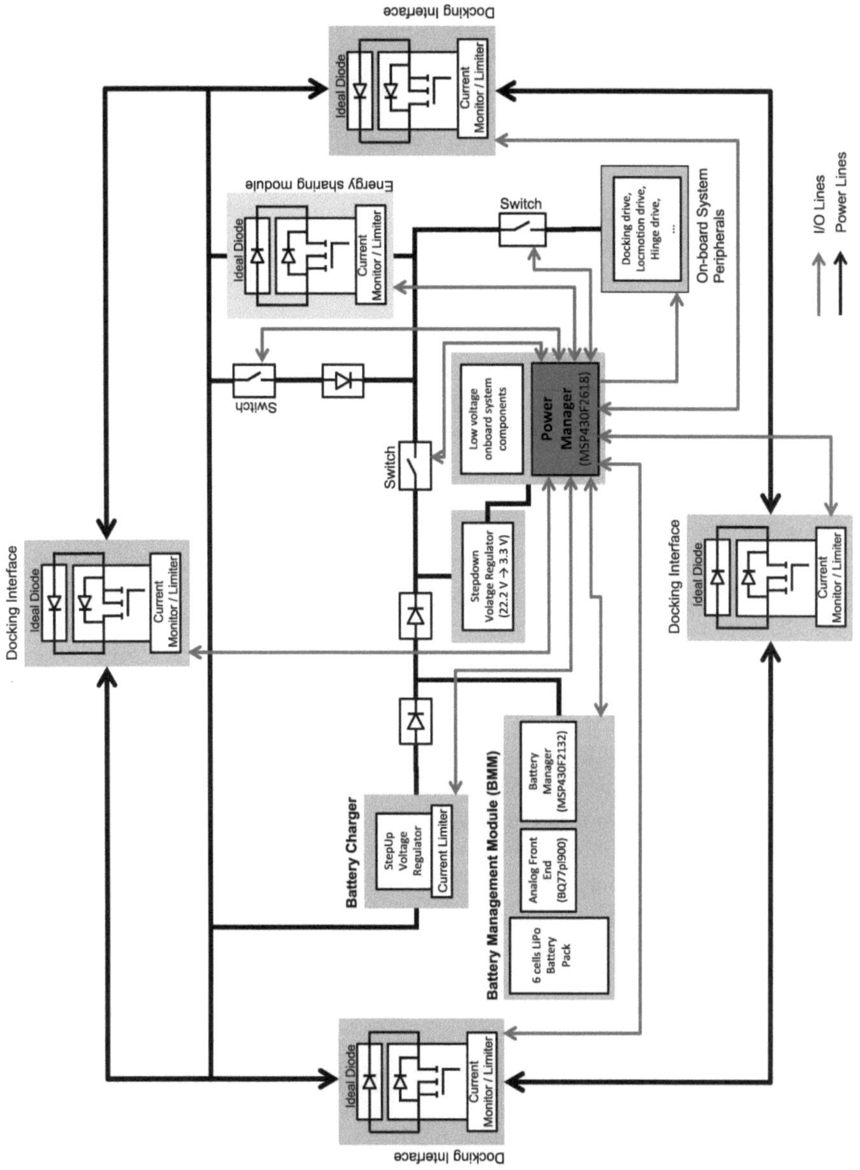

Fig. 4.4 Block diagram of the proposed power management system with fault tolerant energy sharing

Battery management module The custom designed battery management module (BMM) comprises two components: an analogue front end and a battery controller. The analogue front end (AFE) is a 5–10 series cell Lithium-ion battery pack protector, i.e. BQ77pl900, that serves as an AFE in host control mode from TI. It is responsible for managing and controlling the operation of the employed six cells lithium polymer battery pack, for instance, charging and discharging, measuring individual cell voltages, cell balancing, over-charge and discharge protection, etc. The battery controller is a 16-bit ultra-low-power μcontroller, i.e. MSP430F2132, from TI for managing the operations of AFE in host control mode. It is custom programmed to keep track of battery pack charge capacity, the discharging and charging current, individual cell voltage levels during cell balancing, and other safety functions. It also provides an interface to the application layer software components to access and control the operations and capabilities of AFE in a controlled manner.

Power manager The system power manager is a core control module in the power management system. It is a 16-bit ultra-low-power MSP430F2618 μcontroller from TI. It is dedicated to control and monitor the different system functionalities and parameters, respectively, such as, communication with the BMM to extract and control the battery pack parameters, enable or disable various on-board system components, e.g. actuators, sensors, docking switches, energy sharing, etc. Not only that, it also measures the current flow through the energy sharing module and the four docking interfaces. The current flow measurements are then transferred to the core processing unit, i.e. *Blackfin* controller, in order to provide an access to the application layer software routines to control the dynamic current flow collectively through the organism's power bus.

Battery charger The dedicated battery charger is a DC/DC controller that operates as a constant-current and constant-voltage regulator (LT3756 from Linear Technology). It provides two important features in a single chip, namely, the built in boost converter and the current limiter. The boost converter is required to support the concept of *trophallaxis*—recharging an energetically weaker robot by a healthy robot—at hardware layer. And secondly, to protect the battery pack against excess charging current the limiter is required to limit the charging current during trophallaxis. Inside the power management system, in its active state, it acts as a constant current, voltage limited source to the battery pack. To reduce the power losses during trophallaxis, the charge current is limited to 0.5 C, flowing into the battery pack. Here, 1C corresponds to the maximum charge capacity of the cell, i.e. 1,050 mA.

Ideal diode To improve the energetic efficiency of the system, the *ideal diode* — LTC4358 from Linear technology—reduces the power dissipation by replacing a power Schottky diode with an internal 20 Ohm N-channel MOSFET. The device is mainly used because of its low power consumption, low heat dissipation, high current flow, a small foot print, and last but not least, smooth switch-over without oscillations.

Energy sharing module The energy sharing module in the system allows a robotic module to control the outward current flow, i.e. from its on-board battery pack, to other docked robotic modules through the respective docking interface. It is a combination of an ideal diode, a current limiting switch, and a current sensor. The *ideal diode* creates a forward path for the current flowing through the docking interface to the on-board system components. The *current limiter* (CL) not only sets an upper limit on the current flow, also acts as a switch to control the outward current flow. In its *inactive* state, if the on-board battery voltage is lower than the voltage across the power bus, then the power bus current is used to energize the on-board system components. While in its *active* state, the CL in the energy sharing module provides an innate fault tolerance by limiting the outward current flow to 1.9 A.

Docking interface The docking interface is a combination of an ideal diode, a current limiter and a current sensor. On each lateral side of the REPLICATOR robot, the docking interface switch in its default, i.e. *inactive*, state allows a unidirectional current flow into the system. Whereas, in its *active* state, it provides a bidirectional current flow path through the particular docking interface. At each docking interface, the current sensor MAX9928F—uni-/bidirectional, high-side, current-sense amplifier by Maxim—allows the system power manager to measure the current flow, and the current limiter TPS2491—a positive high-voltage, power-limiting hot swap controller by TI—limits the current flow to $\approx$8 A.

4.3 Simulation Framework

Because of the unavailability of sufficient number of REPLICATOR robotic modules at the time of writing this work, a simulation framework was devised to implement and examine the effectiveness of the application layer power management mechanisms. It was especially developed because at that time no such tool or program was available that could be used to perform the desired experiments at a modular robotic organism level. The simulation framework called *Replicator Power Flow Simulator*, was developed in the light of experience gained with the real REPLICATOR robotic modules, presented in Sect. 5.1.

The *Replicator Power Flow Simulator* developed within this research work consists of two parts: the front-end and the simulation engine. The *simulation front-end* was designed in LabWindows/CVI from National Instruments. LabWindows/CVI was chosen, as it facilitates to code the application software layer power flow control strategies, the associated parameters, and in designing the graphical interface to observe and access different measurements and parameters in the organism, respectively. The designed graphical user interface was used to monitor and configure the different parameters affecting the power sharing among the robotic module of an organism. To obtain more realistic simulation results, at the back-end, a SPICE[1] simulation engine from Linear technology, LTSpice IV, was integrated to simulate the power flow between the robotic modules. Figure 4.5 shows the block representation of the designed REPLICATOR Power Flow Simulator.

[1] Simulation Program with Integrated Circuit Emphasis

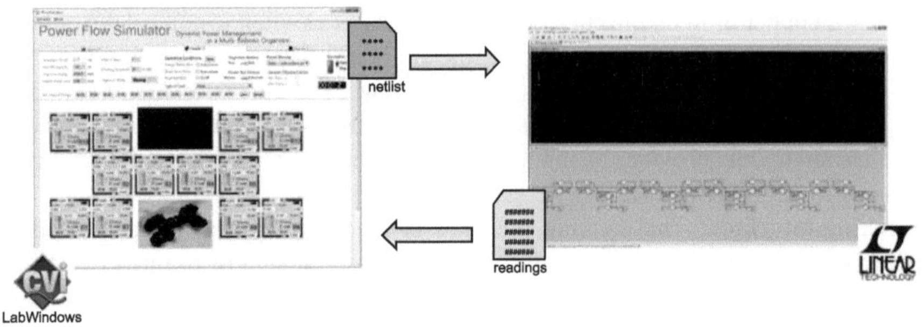

Fig. 4.5 REPLICATOR power flow simulator framework with SPICE simulation engine

To obtain power flow data in the organism, during run-time, i.e. on each iteration, the simulation front-end invokes the simulation engine in DOS mode to process the generated SPICE net-list. The SPICE net-list—*.cir file—contains the power management components and their interconnections that define the power flow between the robotic modules of an organism. The simulation engine processes the net-list and records the current flow data at the specified points in the organism that is then stored on the disk in an ASCII format. The current flow data is then read out to update the variables internally and graphically in the simulation front-end.

4.3.1 System Power Consumption Model

In the simulation framework for the intra-robot and intra-organism power flow, a simple system power consumption model was devised to simulate the electronic load that each robotic module may experience when docked in an organism. The devised system power consumption model at each robotic module distinguishes the system load between *fixed* and *variable load*. The term *fixed* refers to the power consumption of the essential system components, which include core processing unit, peripheral controllers, power management module, and other sensing components, necessary to keep a robotic module alive. Whereas, the *variable load* refers to the power consumption of the actuators, such as, 2D and 3D motor drive, drive controllers, laser sensors, etc., required only during locomotion on different surfaces. The power consumption of these components was therefore considered as *dynamic system load*, since they do not require to be ON all the times, i.e. during the normal operation of a robotic module.

For the sake of simplicity, without considering the complex kinematics involved in the organism motion, the variable load calculation uses the power consumption of a hinge drive used for 3D movement in a REPLICATOR Kabot. Under different load conditions in the organism the power consumption of a hinge drive can be determined from the torque required to lift the "*n*" number of robotic modules, i.e.

$$\tau = (\sum_{i=1}^{n} i) \cdot F \cdot d, \\ = \frac{n(n+1)}{2} \cdot mg \cdot d, \tag{4.1}$$

where the average weight of a REPLICATOR robot is roughly 1 kg, the gravity "g" is $9.8\,\mathrm{m/s^2}$, and the variable "d" is the distance that is moved vertically upward from the pivot point.

The torque "τ_r" required by the actuator motor can be obtained as

$$\tau_r = \frac{\tau}{r_r \cdot \eta}, \tag{4.2}$$

where the gear reduction ratio of the hinge drive is 800 : 1, and the efficiency "η" is roughly set to 75 %. The required current "I_r" with "τ_r" can be therefore obtained as

$$I_r = \frac{\tau_r}{\tau c} + i_0, \tag{4.3}$$

where "τ_c" is the torque constant of the brush-less DC motor, i.e. Portescap 32BF, and equals to 7.8 mNm. The variable "i_0" represents the current consumption of the motor drive without load, which is 65 mA at 12 V.

The motor speed "v_r" required for the desired actuation is then obtained as

$$v_r = v \cdot r_r, \tag{4.4}$$

where "v" is the nominal motor speed and "r_r" is the reduction ratio, as mentioned in (4.2). The voltage required to the motor to produce the desired speed "v_r" can now be obtained as

$$V_r = \frac{v_r}{s_c}, \tag{4.5}$$

where s_c is the speed constant of the brush-less DC motor. Finally, based on the above calculations the power consumption P of the actuator motor can be obtained as

$$P = V_r \cdot I_r. \tag{4.6}$$

4.4 Application Software Layer Power Management

For an efficient utilization of the available energetic resources, an application software layer power management mechanism was devised in order to evaluate the effectiveness of different variants of power sharing during different behavioral states

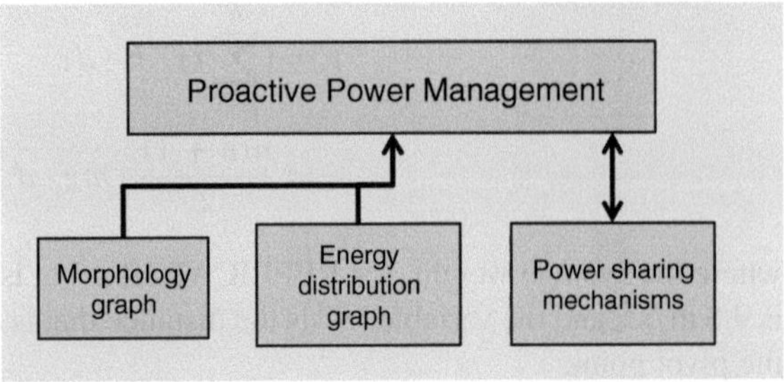

Fig. 4.6 Proposed application software layer power management components of a robotic module

of a modular robotic organism, as defined in Sect. 3.2.2. The power management features, realizable at the application software layer, combine the local and global system information at each robotic module to choose an efficient power sharing topology in a modular robotic organism. Figure 4.6 shows the proposed application software layer power management components of a robotic module. These components include *proactive power management*, *morphology graph*, *energy distribution graph* and *power sharing mechanisms*. Here, in the scope of this work, the implementation details of *proactive power management* are not included.

4.4.1 Morphology Graph

The morphology graph defines the structure of a modular robotic organism. In a real environment, the robotic modules can adapt a pre-programmed or a new morphology, learned from the environment, to overcome the internal or environmental driven challenges, or both. With real robotic modules, prior to an organism formation, a robotic module in the arena may at random choose to become the head module in the organism and broadcast the joining request to its swarm mates.

Figure 4.7a, b show a modular robotic organism comprising 14 robotic modules and the morphology graph, defining its structure, respectively. The organism structure was constructed in Symbricator Robot 3D simulator (Winkler and Wörn 2009). Each node in the morphology graph represents a robotic module and the interconnections show the orientation of the docked robotic modules in the organism. Each robotic module in the organism maintains a copy of the morphology graph that provides it the information about itself and the other docked robotic modules. The entries in the morphology graph are updated with the periodic circulation or transmission of morphology messages by each robotic module in the organism. A *morphology message* structure includes the sender's ID and the IDs of the robotic modules that are its direct neighbors—physically docked. Table 4.6 shows an example morphology message structure.

Fig. 4.7 REPLICATOR modular robotic organism. **a** Modular robotic organism comprising 14 robotic modules, **b** Morphology graph

Table 4.6 An example message structure for maintaining the morphology graph entries in the organism

Message type	Sender's ID	Front link's ID	Right link's ID	Rear link's ID	Left link's ID	Status
MGraph	r2	r1	r3	r7	r5	xx

4.4.2 Energy Distribution Graph

The energy distribution graph is devised to obtain the energetic information of the robotic modules in an organism. It is developed on the basis of morphology graph entries. With the available information, the robotic modules can then tune or synchronize their collective operations, accordingly. For instance, during collective recharging, the robotic modules using the energy distribution graph can decide the recharging order and successively estimate the time required to fully recharge the organism. In another scenario, an organism with an uneven or non-uniform energy distribution may achieve an energetic equilibrium by recharging its energetically weaker robotic modules from the charge of the energetically healthy robotic modules.

Likewise, each robotic module maintains a copy of the energy distribution graph by periodically broadcasting the status messages in the organism. Table 4.7 shows an example message structure. It includes the *sender's ID*, battery pack terminal voltage in volts, battery pack state of charge in percentage, and it's operating mode in the organism—energy *donor*, *acceptor* or *neutral*—as defined in the macroscopic model in Sect. 3.2.2.

Table 4.7 An example message structure for updating energetic status of a robot in the organism

Message type	Sender's ID	Battery terminal voltage (volts)	State of charge (%)	Operating mode
Edist	r5	22.2	80	$dn/acp/neu$

4.4.3 Power Sharing Mechanisms

As identified in Sect. 3.2.2, a modular robotic organism has to adapt different mode of power sharing for its survival in different behavioral states, mainly to accommodate a breakdown situation at a robotic module. For this purpose, following the first part of the artificial energy homeostasis definition, presented in Sect. 3.3.1, the power sharing mechanisms or policies are defined to regulate the power flow between the robotic modules, that can be chosen in different behavioral states of an organism. Utilizing the control architecture of the underlying power management system of a robotic module, described in Sect. 4.2, and the information available from the morphology and energy distribution graphs, following power sharing polices are envisaged at the application software layer of each robotic module.

- **No power sharing**
  This is the most obvious scenario, in which even though having the ability the robotic modules in the organism do not share their on-board energy reserve during their collective collaborative actions in the environment. The robotic modules in this scenario coordinate with each other at the application software layer by sharing sensory information, coordinating tasks, synchronizing movements, etc., but solely depend on their on-board energy reserve to fulfill their power demands. Consequently, following effects can be expected on the behavior of a modular robotic organism without power sharing:
  - During an organism locomotion, the uneven mechanical load distribution creates an imbalance of energy between the robotic modules, in such a situation an energetically weaker robot that runs out of energy can stop the organism to continue its locomotion in the arena.
  - In a similar scenario, the energetically weaker robot(s) in the organism might because of energy deficiency become unable to detach (un-dock) itself (themselves) from other robotic modules or vice versa. In either case, the whole organism becomes paralyzed and requires external assistance.
- **Power sharing without power flow control**
  This power sharing policy, which here is also named *static power sharing*, is characterized by the fact a single power bus is established for power sharing in the organism. The term *static* emphasizes that the organism's power bus topology cannot be changed. In such a configuration, the robotic modules in an organism can share their on-board energy reserves but cannot control the current flow direction through the organism's power bus. The energy donor will always be the

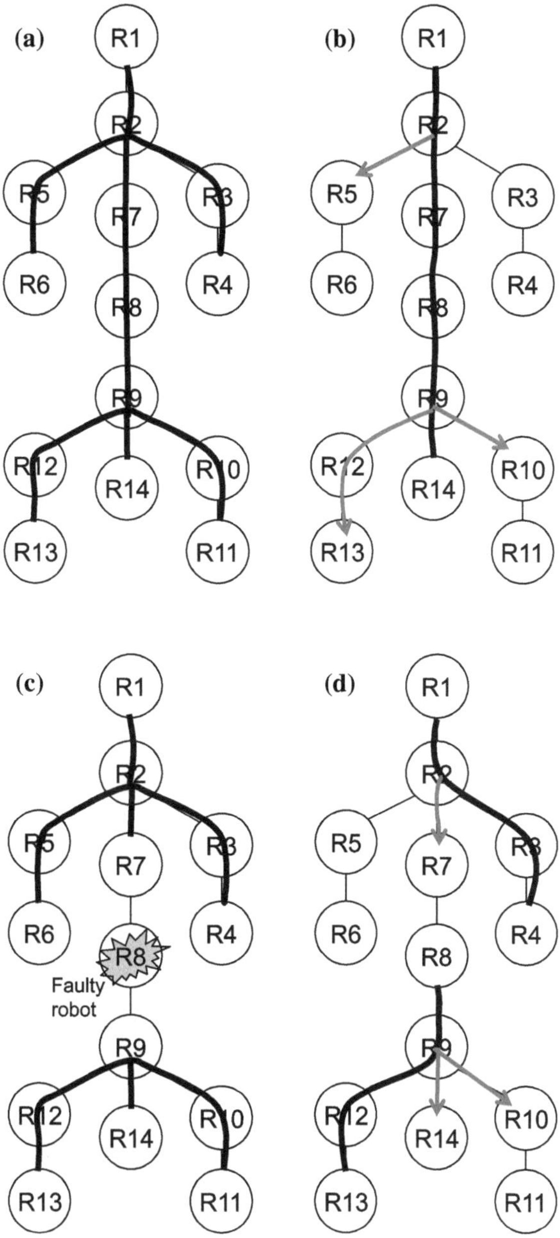

Fig. 4.8 Different forms of power bus topologies in an organism. **a** an omni-directional power bus: power sharing with no control over the power flow between the robotic modules, **b** and **d** power sharing with power flow control between the robotic modules. The *black lines* represent a bi-directional current flow path and the *gray lines* with the arrows represent a uni-directional current flow path in the particular direction, **c** a power sharing scenario with power flow control in the presence of a *faulty* or *broken* robotic module in the organism

energetically stronger robotic module(s), in terms of state of charge, in the organism. Figure 4.8a shows the block diagram of a power sharing scenario with a single organism's power bus that allows an omni-directional current flow between the robotic modules.

Following are the consequences of the described power sharing mechanism that may appear on the operations of a modular robotic organism in different scenarios:

- A drawback of a single organism's power bus appears at the energetically healthy robotic module(s)—energy donors—as they quickly lose their on-board energy reserve. In addition, depending on the number of dependent and energy donor robotic modules in an organism, an omni-directional power bus may induce high current flow between the robotic modules that may decrease system efficiency because of the increased power losses in the system.
- Consider a situation in which a fault, e.g. a short circuit, either due to a mechanical or an electronic failure occurred in a robotic module. In a worst-case scenario, since the robotic modules cannot control the current flow direction through the organism's power bus, the whole system may collapse, i.e. the whole electronic circuitry may get damaged.

• **Power sharing with power flow control**
This form of dynamic power management is designed to overcome the short comings of the above mentioned power management mechanisms. Following the artificial energy homeostasis definition, with this form of power management, the robotic modules in the organism can control the direction of current flowing through the organism's power bus. In other words, the robotic modules can form more than one or sub power buses in the organism, as required. Figure 4.8b, d show the two power sharing scenarios with power flow control between the robotic modules. With having control over the organism's power bus, the robotic modules can isolate the system malfunctions or abrupt system failures, e.g. a short circuit, sensor and actuator failures, etc., individually and collectively in the organism. The isolation of a faulty component or a robotic module(s) in turn allows the rest of the robotic modules in the organism to restore their normal operations. Figure 4.8c shows an example scenario, in which the robotic modules despite a faulty robot can still share their on-board energy reserves with each other by forming sub-power buses in the organism.

  - ***Power sharing policy 1***
  This form of power sharing is devised on the *local energy distribution* information available to each robotic module in the organism. In this topology, every robotic module in the organism using the morphology and energy distribution graph maintains a *local energy distribution table*. The table includes its state of charge (SOC) and the battery SOC of its direct neighbors, docked to its four docking interfaces. Table 4.8 shows the local energy distribution table, periodically updated at each robotic module in the organism.
  An energetically healthy robotic module, i.e. having $SOC_{robot_{(i)}} > E_{th}$, shares its on-board energy reserve if anyone among its direct neighbor's SOC is less than E_{th}—energy sharing threshold. In such a scenario, an uni-directional power flow path between energy donor and acceptor is established by activating the electronic gates on the respective docking interfaces.

Table 4.8 Energy distribution table with power sharing policy 1

Myself	Front neighbor	Right neighbor	Rear neighbor	Left neighbor
SOC_{robotX}	$SOC_{robotX+1}$	$SOC_{robotX+2}$	$SOC_{robotX+3}$	$SOC_{robotX+4}$

Table 4.9 Energy distribution table with power sharing policy 2

	Front neighbor	Right neighbor	Rear neighbor	Left neighbor
SOC_{robotX}	$SOC_{robotX+1}$	$SOC_{robotX+2}$	$SOC_{robotX+3}$	$SOC_{robotX+4}$
$SOC_{robotX+3}$	SOC_{robotX}	$SOC_{robotX+11}$	$SOC_{robotX+12}$	$SOC_{robotX+13}$
$SOC_{robotX+4}$	$SOC_{robotX+16}$	SOC_{robotX}	$SOC_{robotX+14}$	$SOC_{robotX+15}$
$SOC_{robotX+1}$	$SOC_{robotX+6}$	$SOC_{robotX+7}$	SOC_{robotX}	$SOC_{robotX+5}$
$SOC_{robotX+2}$	$SOC_{robotX+8}$	$SOC_{robotX+9}$	$SOC_{robotX+10}$	SOC_{robotX}

- ***Power sharing policy 2***
  This form of power sharing uses the *global energy distribution* information available to each robotic module in the organism. In this topology, each robotic module in the organism maintains a global energy distribution table that not only keeps record of its energetic status and the direct neighbors but also the direct neighbors of its direct neighbors, as well. Table 4.9 shows the entries in the global energy distribution table. An energetically healthy robotic module in the organism shares its on-board energy reserve if anyone among its direct neighbors or their direct neighbor has SOC less than E_{th}. The energy donor robotic module in such a scenario establishes a power bus between itself and the acceptor robotic modules by closing the electronic switches present in a particular path, at the respective docking interfaces.

In the context of this research work, the morphology and the energy distribution graphs along with the power management policies are implemented at each robotic module in the *Replicator Power Flow Simulator* to explore the energetic behavior of two exemplary modular robotic organisms, see Sects. 5.2 and 5.3.

4.5 Application Software Layer Fault Tolerance

The long term physical aggression of robotic modules in an organism that depends on the level of coordination and collaboration each robotic platform owns, demands a robust system design that can tolerate the presence of faults and components failures. In other words, the fault tolerance of a modular robotic organism depends on the mutual ability of the robotic modules to react in response to intra-organism malfunctions. Faults at an individual level in an organism generally originate either due to *endogenous* or *exogenous* component malfunctions, also mentioned by Isermann and

Ballè (1997). The conceived application software layer fault tolerance in a modular robotic organism implements the fault detection and identification and fault isolation algorithms.

4.5.1 Fault Identification

Before focusing on fault tolerance mechanisms at an organism level, it is essential to identify the different types of faults that can halt or cease the collaboration of multiple robotic modules. Figure 4.9 shows the fault tree of a modular robotic organism using a fault tree analysis approach. The operations in a modular robotic organism can become irregular mainly because of a robot failure, organism's power bus failure, communication failure or some external influence in the from of a mechanical damage. The causes of a *robot failure* are covered in detail in Appendix A. The *organism's power bus* failures are mainly linked with the docking units that usually arise because of a misalignment between the docking interfaces of the robotic modules or a defective docking connection, highlighted in figure A.6, (see Appendix A).

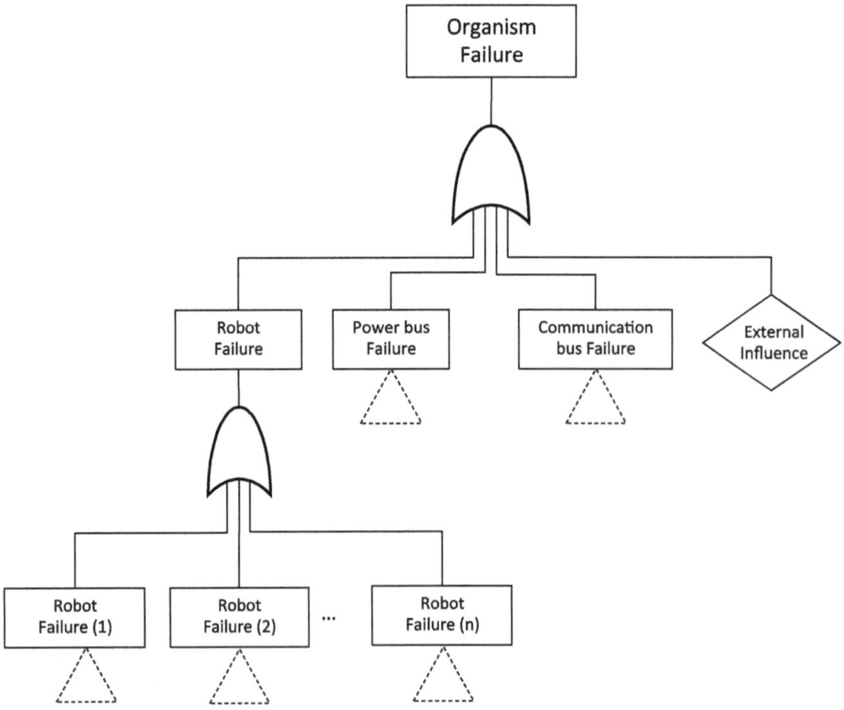

Fig. 4.9 Fault tree combining the probable failure causes behind an organism failure

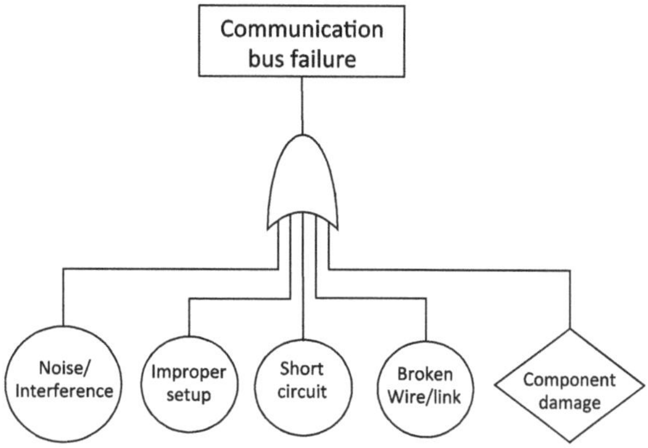

Fig. 4.10 Organism's communication bus failure fault tree

- **Communication bus failure**
  A communication medium, both wired and wireless, between the robotic modules in an organism is essential to establish and sustain their collaboration and coordination. As described in Sect. 4.1.4, a wired communication link between the docked REPLICATOR robotic modules is established by means of contact pins at the docking interfaces. In such a configuration, a communication bus failure in a modular robotic organism can develop mainly because of noise or electrical interference, an improper configuration of communication modules, a short circuit on the communication path, a broken wired link, or a component damage. Figure 4.10 shows the fault tree combining the probable failure causes behind an organism's communication bus failure. The electrical interference due to fluctuating voltage levels, flux from the voltage regulators, etc., may introduce random noise in the sensor data or in the communication channels.

Considering the design of a modular robotic organism, the impact of a fault or system failure on multiple robotic modules mainly depends on three factors: the *fault type*, the *position* and *role*—responsibility—of the faulty robotic module in the organism. The *fault type* determines the severity of the following consequences. The *position* of a faulty robot in the organism becomes critical as it determines the number of directly and indirectly affected robotic modules. And finally, its *role* in the organism from the perspective of power sharing becomes relevant as it may affect the operations of energetically healthier and weaker robotic modules.

4.5.2 Fault Detection and Identification Algorithm

The tolerance against abrupt faults depends on a system's instantaneous reaction upon the detection of an anomaly affecting its equilibrium. For the detection of endogenous and exogenous faults, their location and isolation, the homeostatic control of a robotic module, described in Sect. 3.3.1 regulates its essential variables by continuously monitoring the signals from the four docking interfaces, the energy sharing module and the battery management module.

Algorithm 4.5.1: APPLICATION LAYER FAULT DETECTION AND IDENTIFICATION()

```
global X := robot's id
global Fault_type := UNKNOWN
global Fault_detected := false
global Fault_location := UNKNOWN

comment: Fault detection check at the energy sharing module
if r(X)(Esh) := ENABLE and CL_r(X)(Esh)[PwrGD] == low
  then
    for l ← 1 to L
      do
        if (I_sh_r(X)[(n − N), ..., (n − 1), (n)] == Fault_Signature(l))
          then
            r(X)→(Esh) : = DISABLE
            for each dkInt := {Front, Right, Rear, Left}
              do {r(X)→(dkInt) := DISABLE
            Fault_type := KNOWN
            Fault_detected := true
            Fault_location := UNKNOWN
            Broadcast(r(X) → faulty robot)
            exit
    Fault_detected := true
    Fault_type := UNKNOWN
    Fault_location := UNKNOWN
  else
    comment: Fault detection check at the four docking interfaces
    for each dkInt := {Front, Right, Rear, Left}
      do
        if r(X)(dkInt) := ENABLE and CL_r(X)(dkInt)[PwrGD] == low
          then
            for l ← 1 to L
              do
                if (I_dkInt_r(X)[(n − N), ..., (n − 1), (n)] == Fault_Signature(l))
                  then
                    r(X)→(dkInt) := DISABLE
                    Fault_type := KNOWN
                    Fault_detected := true
                    Fault_location := r(X)→(dkInt)
                    Broadcast(r(X)→(dkInt) := faulty robot)
                    exit and iterate on other docking interfaces
            comment: Fault detected but its type and source is still unknown
            Fault_detected := true
            Fault_type := UNKNOWN
            Fault_location := r(X)→(dkInt)
        else if r(X)(dkInt) := DISABLE
          then
            for l ← 1 to L
              do if I_dkInt_r(X)[(n − N), ..., (n − 1), (n)] == Fault_Signature(l)
                then
                  Fault_type := KNOWN
                  Fault_detected := true
                  Fault_location := r(X)→(dkInt)
```

In the designed framework, for a fault detection and identification, the homeostatic control of a robotic module periodically samples the current flow through each docking interface, energy sharing module and the battery management module. In this regard, Algorithm 4.5.1 was devised to detect and identify the possible faults and failures in the power managements system. The first "if" statement in the Algorithm 4.5.1 is meant to monitor the status of an energy sharing module in the robot, i.e. the power good (*PwrGD*) signal when it is enabled. In the case of an anomaly—low *PwrGD* signal from the energy sharing module during its enabled/active state—the current flow values through the energy sharing module are matched with a predefined fault signature, e.g. using pattern matching. The *fault signatures* can be obtained in different operating conditions to classify between different types of faults in regard to the power flow in the organism. On the detection of a fault, the energy sharing through the particular robotic module is stopped instantly to find its location afterwards. In addition, the faulty robot broadcasts its faulty status in the organism. In the case the energy sharing module is operating in its normal state, it monitors the current flow through the four docking interfaces of the particular robotic module to detect and identify the exogenous faults.

4.5.3 Fault Isolation Algorithm

After the detection of an occurred fault in the system, a similar procedure is then adapted for the isolation of the fault at each robotic module in the system. In the case

Algorithm 4.5.2: APPLICATION LAYER FAULT ISOLATION()

```
if Fault_detected == true and Fault_location == UNKNOWN
then {
    comment: locating if an endogenous fault has occurred
    for each dkInt := {Front, Right, Rear, Left}
      do r(X)→(dkInt) := DISABLE
    comment: Now, activate energy sharing
    r(X)→(Esh) : = ENABLE
    for l ← 1 to L
      then { if I_sh_r(X)[(n − N), ..., (n − 1), (n)] == Fault_Signature(l)
               then { r(X)→(Esh) : = DISABLE
                      Fault_type := KNOWN
                      Fault_location := ENDOGENOUS
                      Broadcast(r(X)→faulty robot)
                      exit
    comment: Identifying an exogenous fault source
    for each dkInt := {Front, Right, Rear, Left}
      do { r(X)→(dkInt) := ENABLE
           for l ← 1 to L
             do { if I_sh_r(X)[(n − N), ..., (n − 1), (n)] == Fault_Signature(l)
                    then { r(X)→(dkInt) := DISABLE
                           Fault_type := KNOWN
                           Fault_location := EXOGENOUS
                           Broadcast(r(X)→(dkInt)→faulty robot)
                           exit from inner FOR loop and scan other docking interfaces
```

a fault is detected with an unknown location, the Algorithm 4.5.2 implements a simple mechanism to locate and isolate a faulty robot in a modular robotic organism. Firstly, the algorithm tries to locate a fault as if it has occurred within the system. For this purpose, it opens the current flow path through its four docking interfaces and enables the energy sharing. In this state, if the current flow measurements through the energy sharing module are matched with a fault signature, then the fault location is declared as "endogenous". In such a condition, it then disables the energy sharing and the four docking interfaces of the particular robotic module and broadcasts its faulty status in the organism. So that, the docked robotic modules can adapt their behavior accordingly. In the case no fault has occurred within its system, it scans the current flow measurements through each docking interface, one after the other to detect an *exogenous* fault situation. In the presence of a fault situation at a docking interface, the algorithm disables (deactivate) the particular docking interface and broadcasts an alert message in the organism, i.e. the presence of a faulty robot on the particular docking interface. The broadcast message may in turn help the remaining robotic modules, that are not directly docked to the faulty robot, in adapting their behavior accordingly.

4.6 Summary

This chapter covered the implementation details of the presented concept to address the self-sufficiency problem at a modular robotic organism level. In the beginning, it described the hardware design considerations to highlight the factors involved in the design of a dynamic power management system, i.e. a robotic module's electronic architecture, power budget calculations, choice of system voltage, and platform specific constraints. With all such considerations, it then presented the hardware design of the proposed dynamic power management system of a robotic module. The characteristic features of the proposed power management system, beneficial for the dynamic operations of a modular robotic organism, are the dynamically controllable power sharing and innate fault tolerance. In addition, the architecture allows to access and control the behavior of the hardware components of a robotic module at the application software layer to coordinate its actions with the docked robotic modules. In the later half of the chapter, for the implementation of application software features, it presented the devised simulation framework that was solely developed to conduct the power sharing experiments at an organism level. In succession, the implementation details of application layer power management and fault tolerance are extensively covered.

Chapter 5
Experiments and Results

Abstract This chapter presents the results of the experiments conducted to explore the behavior of a modular robotic organism with the proposed hardware and the application software features in the simulation framework. In the first half, the chapter presents the experimental results obtained after integrating the proposed power management system in the hardware design of the real robotic modules during artificial trophallaxis, power sharing and an emulated fault condition. In the second half of the chapter, the application software, implemented in the simulation framework, is tested by simulations. The simulation experiments are broadly divided into two parts: power sharing and fault tolerance. The power sharing simulation experiments include the simulations of dynamic power sharing in two exemplary organism morphologies. The fault tolerance simulation experiments include the results obtained by adapting the procedure defined in the fault detection and identification and fault isolation algorithms, during different fault situations. At the end, a brief summary concludes the chapter.

5.1 Experiments with the Proposed Hardware

In the first place, the proposed power management system's behavior, integrated in the hardware design of REPLICATOR robotic modules, was explored in three scenarios: artificial trophallaxis, power sharing and fault tolerance.

5.1.1 Artificial Trophallaxis

The artificial trophallaxis feature allows the energetically healthy robotic modules to donate a portion of their on-board energy reserve to energetically weaker robotic modules in the arena. The energy transfer begins after a successful docking of two robotic modules—energy donor and acceptor—by recharging the energetically

R.H. Qadir, *Self-Sufficiency of an Autonomous Reconfigurable Modular Robotic Organism*, Adaptation, Learning, and Optimization 17,
DOI 10.1007/978-3-319-10289-4_5

Fig. 5.1 Artificial trophallaxis test setup

weaker robot's battery pack. The efficiency of food exchange in natural systems may not be concerned due to several reasons. But in an artificial system, with rechargeable battery packs, its efficiency cannot be ignored.

For efficiency measure, the artificial trophallaxis was emulated using two battery packs and a battery recharging module. Figure 5.1 shows the test scenario in which a battery with low state of charge (SOC) was recharged from a battery pack with relatively high SOC. The trophallaxis efficiency—η_{troph}—is obtained as

$$\eta_{troph} = \frac{\text{charge taken out from the acceptor robot's battery pack after recharging}}{\text{charge delivered by the donor robot during recharging}} \cdot 100\,\%$$

$$= \frac{1}{k} \cdot \sum_{k} \frac{C_{acpt}(k)}{C_{dnr}(k)} \cdot 100\,\%, \tag{5.1}$$

that is, the amount of charge taken out from the acceptor (energetically weaker) robot's battery pack (after recharging) divided by the amount of charge taken from the donor robot's battery pack during recharging, where k denotes the discrete time instance.

Such a set-up was purposefully chosen to solely measure the trophallaxis efficiency, since it does not include the power consumption of the on-board electronics on both of the robotic modules. Figure 5.2 shows the measurements obtained during energy transfer between the two battery packs. Figure 5.2a, c show the donor robot's and the acceptor robot's battery pack voltages during trophallaxis, respectively. And, Fig. 5.2b, d show the current flow from the donor robot's battery pack to acceptor robot's battery pack, respectively. During energy transfer the donor battery voltage decreased as the acceptor battery voltage increased. Correspondingly, the current flow from the donor battery increased as the terminal voltage dropped. On the other side, the recharging module recharged the acceptor battery with a constant current.

During trophallaxis, the energy was transferred between the two battery packs for a period of about 70 min. Measuring the SOC of the acceptor's robot battery pack, the trophallaxis efficiency obtained using Eq. (5.1), varied between 82–85 %. As already mentioned, the efficiency of the process will decrease in the real robots, as along with battery recharging the on-board electronics of the acceptor robot also consumes power during energy transfer.

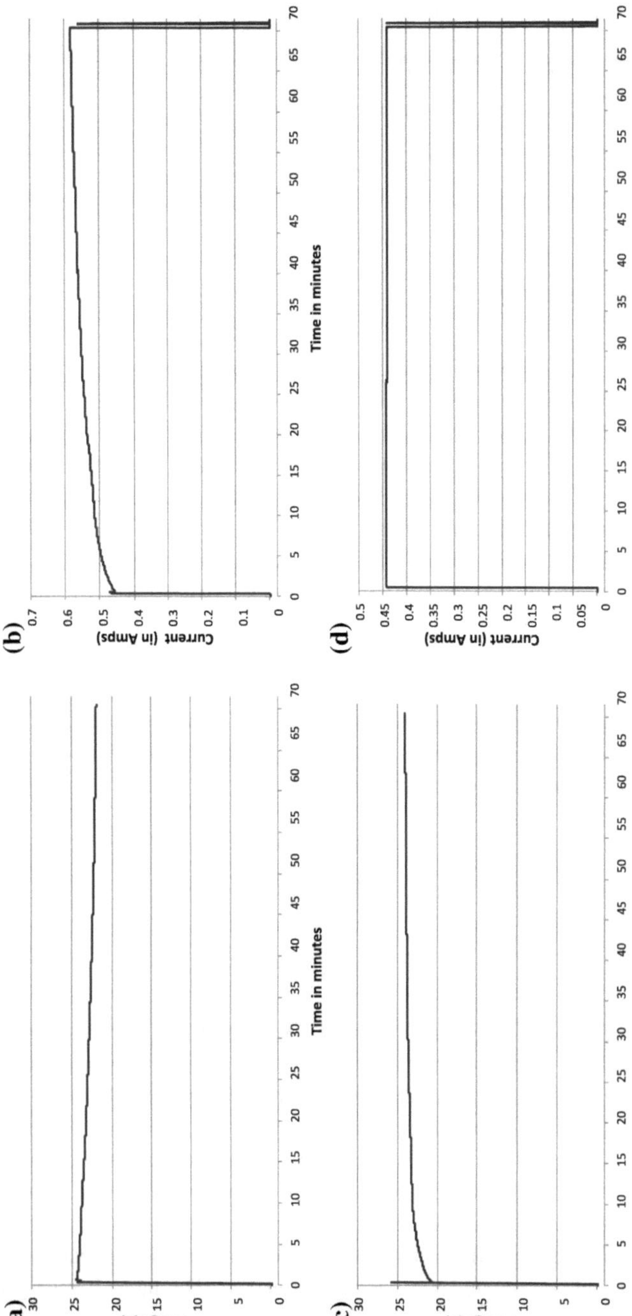

Fig. 5.2 Artificial trophallaxis, **a** energy donor robot's battery pack voltage during energy transfer, **b** current flow out of the donor robot's battery pack, **c** energy acceptor robot's battery pack voltage during energy transfer, **d** current flow into the acceptor robot's battery pack

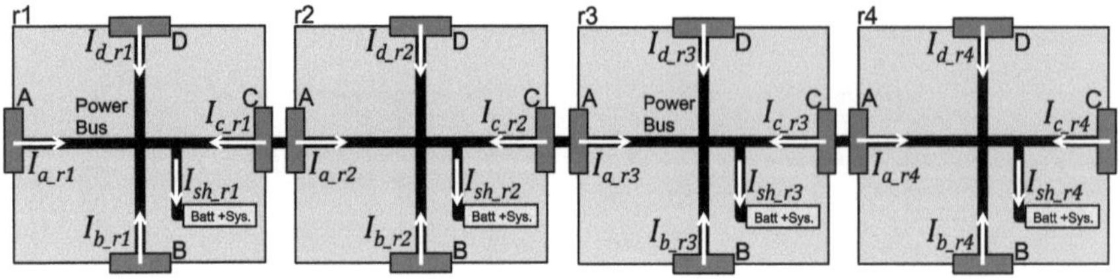

Fig. 5.3 Block diagram representation of the organism's power bus between four robotic modules

5.1.2 Power Sharing

Power sharing is an essential feature of robotic modules that can physically dock with their teammates to achieve common objectives that are beyond the capabilities of an individual robotic module. It allows the autonomous robotic modules to sustain their physical collaboration for a longer period of time. The power sharing experiment between multiple robotic modules was performed to measure and analyze their collective behavior in different scenarios.

Figure 5.3 shows the block diagram representation of a power bus established between four physically docked robotic modules. At each robotic individual, the system *power manager* measures the inward and outward current flow through the energy sharing module and the four docking sides. Consider $I_{a_{r1}}$, $I_{b_{r1}}$, $I_{c_{r1}}$, $I_{d_{r1}}$, and $I_{E_{r1}}$ are the variables representing the current flow through the four docking interfaces and the energy sharing module of a robot r1, respectively.

From Fig. 5.3, using Kirchhoff's current law, the current flowing through the four sides of robot $r1$ can be obtained as

$$I_{a_{r1}} = I_{E_{r1}} - (I_{b_{r1}} + I_{c_{r1}} + I_{d_{r1}}), \tag{5.2}$$

$$I_{b_{r1}} = I_{E_{r1}} - (I_{a_{r1}} + I_{c_{r1}} + I_{d_{r1}}), \tag{5.3}$$

$$I_{c_{r1}} = I_{E_{r1}} - (I_{a_{r1}} + I_{b_{r1}} + I_{d_{r1}}), \tag{5.4}$$

$$I_{d_{r1}} = I_{E_{r1}} - (I_{a_{r1}} + I_{b_{r1}} + I_{c_{r1}}), \tag{5.5}$$

where $I_{E_{r1}} = I_{Batt_{r1}} - I_{Sys_{r1}}$. The variable $I_{Batt_{r1}}$ and $I_{Sys_{r1}}$ represents the current flow from the on-board battery pack and to the system electronics, respectively. Similarly, the current flow from other robotic modules can be obtained.

Figure 5.4 shows the experimental setup used to test the power sharing between four REPLICATOR Kabot modules, docked in a chain formation. For the controlled power flow between the robotic modules, each robotic module was connected with an external power source that also served the purpose of electronic load in the system. The current measurements from each robotic module were periodically sampled and transferred to a PC connected via a UART interface. The later discussion uses the robots numbering convention shown in Fig. 5.3.

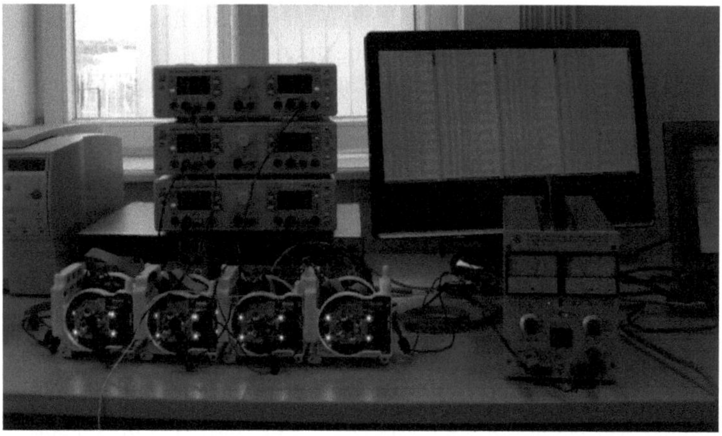

Fig. 5.4 Experimental setup used to test the power sharing between four REPLICATOR Kabot modules

For dynamic power sharing between the robotic modules, robot r4 in the organism through its docking side C was connected to an external power source that can deliver a continuous current of 8 A at 25 V. In addition, an external power supply was also connected at its battery terminals that can deliver a maximum of 2 A of continuous current at 25 V. Robot r3 docked to r4 at side C, was connected to a single power source at its docking side B. Similarly, robot r2 docked to r3 at its side C in the chain had a single power source connected at its docking side B. And lastly, robot r1 docked to r2 through its docking side C was connected to two power sources, at its docking side B and side A, respectively. It is important to consider that the robots r3, r2 and r1 lack a power source at their battery terminals. This implies, the on-board electronics of these robots solely depends on external power connected at any one of their docking sides. The current flow between the robotic modules started as the voltage levels on the power supplies varied with respect to each other. The plots in Fig. 5.5a, b, c and d show the current flow measurements through docking sides A, B, C and the energy sharing module of robot r4, r3, r2, and r1, respectively. The positive current values in the plots indicate an inward current flow through the particular docking interface or the energy sharing module, and the negative current values indicate an outward current flow through the particular component.

Varying the voltage levels at the power supplies connected to the robotic modules induced current flow through the organism's power bus. For a detail analysis, consider the current flow measurements through robot r4 in the organism. In the beginning, an inward current of $\approx$600 mA flows through side C of robot r4. Since the on-board source voltage was greater than the power bus voltage, an outward current flow of $\approx$100 mA was measured through its energy sharing module. The current flow through side B remained zero, while the inward current through side C then left the system through side A, connected to robot r3. Decreasing the voltage level on the power supply connected to side C of robot r4 in comparison with the voltage source connected to its battery terminals increased the current flow through the energy sharing module after the 16th minute on the time scale.

From the current flow measurements at robot r3, an inward current flow into the system through side C, connected to side A of robot r4, was measured. Since, no power source was connected at the battery terminals of robot r3, the power bus current through the energy sharing module energized the on-board electronics. This is the reason, the current flow through its energy sharing module remained negative. In proportion to the voltage levels on the two power supplies present at side A and side B of robot r3, the inward current flow through docking side C left the system through its docking side B and side A.

At robot r2, because of the absence of a power source at the battery terminals, the inward current flow through docking side C energized the on-board electronics. The current flow measurements through the energy sharing module and the docking sides of robot r2 can be seen in Fig. 5.5c. Similarly, in proportion to the voltage levels on the power supplies connected directly and indirectly to its three docking sides A, B and C, the inward current flow through side C left through side A and side B. And finally, from the measurements shown in Fig. 5.5d, a similar behavior at robot r1 in the organism was observed.

The current flow measurements through the organism's power bus show the collective behavior of the robotic modules during power sharing. In addition, the measurements provide the current limit, i.e., maximum current, that a docking interface and energy sharing module of a robotic module permits. The power sharing measurements show that the docking interface current limiter allows a maximum of $\approx$7.2 A of continuous current at the nominal system voltage to flow through, while the current limiter at the energy sharing module limits the outward current flow to $\approx$1.9 A.

In the absence of an on-board power supply at robot r3, r2 and r1, the designed power management system allowed the robotic modules to energize their on-board system electronics from the organism's power bus. The particular feature of the designed power management system maintains operation of those robotic modules that are energetically weak or dependent on other robotic modules in an organism.

5.1.3 Innate Fault Tolerance

Not only to assure the robustness but also from the safety and security perspective of any system, the system's ability to withstand abrupt failures can never be neglected. For example, an ordinary toaster will always be equipped with a fuse that instantly blows in case of a short circuit.

Considering the nature of collaboration between the robotic modules of an organism, the proposed power management system with fault tolerant energy sharing includes several current limiters and current sensing components to add resistance against potential system faults and failures. The current limiters in the system impose an upper limit on the current flow in a particular direction and the current sensor provides a measure of current flow in either direction. The fault tolerance test was

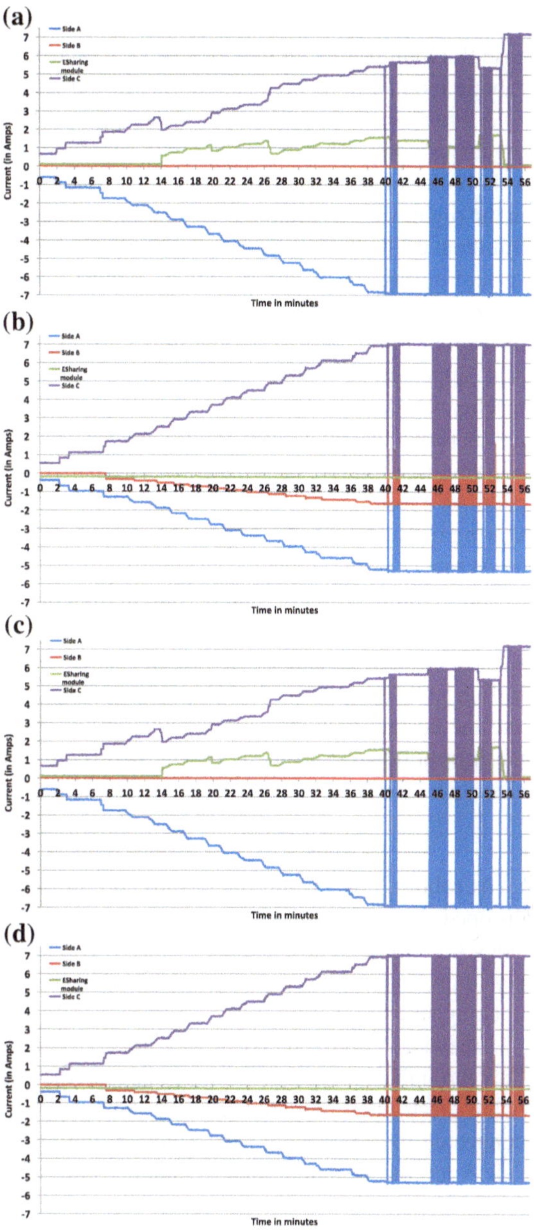

Fig. 5.5 Power sharing in a REPLICATOR organism **a** current flow through the docking interface A, B, C and energy sharing module of robot r4, **b** current flow through the docking interface A, B, C and energy sharing module of robot r3, **c** current flow through the docking interface A, B, C and energy sharing module of robot r2, **d** current flow through the docking interface A, B, C and energy sharing module of robot r1

conducted during power sharing to observe the collective behavior of individuals, in case if a fault occurs on the organism's power bus. For example, a short circuit in a robotic module of an organism during power sharing.

A *short circuit* in a robotic module induces a high inrush current flow through the organism's power bus from the donor robotic modules in the organism. To emulate a short circuit in a running system, the external load applied to the robotic modules in the organism was gradually increased in order to reach the current limit set by the limiter at each docking side. To highlight the fault tolerant behavior of the proposed system during power sharing in an organism, the measurements shown in Fig. 5.5 are zoomed in Fig. 5.6a, b, c and d. The plots in Fig. 5.6a, b, c and d show the measurements recorded at robot r4, r3, r2, r1, respectively. The oscillations in the plots appeared as soon as the current flow through the docking side A of robot r4 reached the threshold limit set by the current limiter. In reaction, the current limiter on the respective docking side disconnected the current flow path and on the expiry of a waiting timer, set in the hardware, tried to re-establish the particular current flow path. As a result, the inward current flow through its docking side C and the energy sharing module instantly dropped to zero.

At robot r3, consequently, the inward current flow through docking side C dropped to zero. And in response, the power supply connected to the docking side B being higher in voltage supplied the power to the on-board electronics and the robotic modules connected at docking side A. As can be seen from the measurements shown in Fig. 5.6b, the current flow through docking side B of robot r3 became positive and the current flow through docking side A dropped to the amount equivalent to the current flown in through docking side B minus the current flown to the on-board electronics, i.e.,

$$I_{A_{r3}} = I_{B_{r3}} - I_{sys_{r3}}. \tag{5.6}$$

At robot r2, as a result the inward current flow though docking side C instantly dropped from 5 A to approx. 1 A. In response, the current flow through the docking sides A and B also dropped, accordingly. But notably, an uninterrupted power supply to the on-board electronics continued from the inward current flow through the docking side C.

Likewise, a similar behavior was observed at robot r1 in the organism. The current flow through docking side C was dropped to the amount required to energize the on-board electronics. While, due to the voltage difference on the two power supplies connected to the docking side A and B, the inward current flow through docking side A left the system though its docking side B, i.e.,

$$I_{A_{r1}} \approx I_{B_{r1}}. \tag{5.7}$$

The above analysis of the current flow measurements during power sharing shows the fault tolerant behavior of the proposed system that without any dedicated control mechanism allowed the robotic modules to continue power sharing despite a fault condition at a robotic module in the organism. The design of the proposed system can be used as a base platform to further develop the adaptive power sharing and fault

Fig. 5.6 Innate fault tolerance feature of the power management system: each graph shows the current flow through the respective robot in the organism. **a** robot r4, **b** robot r3, **c** robot r2, **d** robot r1

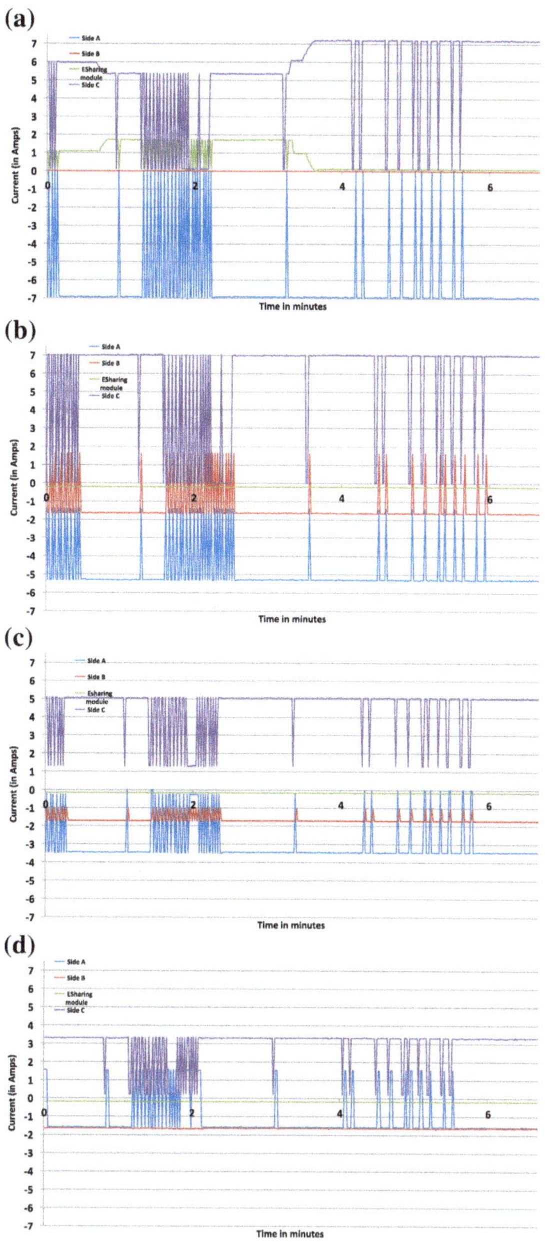

tolerance mechanisms at the application software layer to achieve self-sufficiency in different organism morphologies.

5.2 Simulating Power Sharing in Organism 1

To start with a simple scenario, a REPLICATOR organism consisting of six Kabot modules docked to each other in a chain formation was chosen. Figure 5.7 shows the REPLICATOR modular robotic organism of type 1. The organism structure was constructed in Symbricator Robot 3D simulator (Winkler and Wörn 2009). Figure 5.7a shows the organism's locomotion on a flat surface exhibiting a sinusoidal movement pattern, and in Fig. 5.7b the organism is shown while crossing an obstacle on an uneven surface.

5.2.1 Kinematic Model

For the simulation of dynamic power flow through the organism's power bus, a simple hypothetical kinematic model based on the sinusoidal movement pattern of the robotic modules was devised. The kinematic model captures the movement pattern of the robotic modules in an organism while performing locomotion in the arena. At each motion step, the movement pattern of the robotic modules was used to obtain the power demand, i.e., the power required to perform the particular motion step, at each robotic module in the organism. During simulation, based on the active power sharing policy in the organism the obtained *power demand* was then subtracted from the respective robotic module's or the donor robot's battery pack.

Since it was quite complex and difficult to mathematically model the kinematics of such a modular robotic organism that involves multiple degrees of freedom, also mentioned by I-Ming et al. (1999), therefore for the sake of simplicity the matrix notation was opted to define the motion pattern of the robotic modules in the organism. The particular kinematic model uses the 3D hinge drive on each robotic module to

Fig. 5.7 Modular robotic organism of type 1 consisting of 6 REPLICATOR Kabot modules, **a** while moving on a flat surface, **b** while crossing an obstacle on an uneven surface

model the collective locomotion of the robotic modules in organism 1. Figure 5.8 shows the graphical view of the sinusoidal movement patterns of the robotic modules in organism 1. Variables $J0, \ldots, J5$ represent the movable joints between the robotic modules. On each iteration, the joints between the robotic modules were moved synchronously either vertically upward or downward to achieve a desired motion pattern in the organism. After motion pattern 6, the organism adapts the motion pattern 3–1 for the completion of next motion cycle, as highlight in the Fig. 5.8.

The motion matrix, MM_{Org1}, was then obtained from the movement patterns of the movable joints in organism 1. Each row in the motion matrix represents a motion pattern. Where, “0” in the matrix represents *no motion*, a *positive value* represents the motion in upward direction from the pivot position and a *negative value* represents the backward movement—back to the pivot

$$MM_{Org1} = \begin{bmatrix} 0 & 0 & 0 & 0 & 0 & 0 \\ 1 & 2 & 0 & 0 & 0 & 0 \\ 0 & 0 & 3 & 0 & 0 & 0 \\ 0 & -1 & 0 & 0 & 0 & 0 \\ -1 & 2 & 0 & 0 & 0 & 0 \\ 0 & 0 & 0 & 3 & 0 & 0 \\ 0 & 0 & 0 & -2 & 0 & 0 \\ 0 & -1 & 2 & 0 & 0 & 0 \\ 1 & 0 & 0 & 0 & 2 & 0 \\ 0 & 0 & 0 & -2 & 0 & 0 \\ 0 & 2 & -1 & 3 & 0 & 0 \\ 0 & -2 & 0 & 0 & 0 & 1 \\ -1 & 2 & 0 & 0 & -2 & 0 \\ 0 & 0 & 3 & -1 & 2 & 0 \\ 0 & 0 & 0 & 3 & 0 & -1 \\ 0 & 0 & -1 & 0 & -2 & 0 \\ 0 & -1 & 3 & 0 & 0 & 0 \end{bmatrix}$$

position. Further more, the numbers in the column represent the mechanical load on a particular robotic module on every motion step. That is, the number of robotic modules that are required to be lifted vertically upward or downward by a particular robotic module in the organism.

5.2.2 Simulation Setup

The dynamic power flow between the robotic modules of an organism was simulated on the basis of two significantly important parameters. It includes the *initial energy distribution* among the robotic modules and the *mechanical load distribution* associated with the movement of the organism on a particular terrain. Table 5.1 shows the test scenarios devised for organism 1 with respect to the *initial energy* and *mechan-*

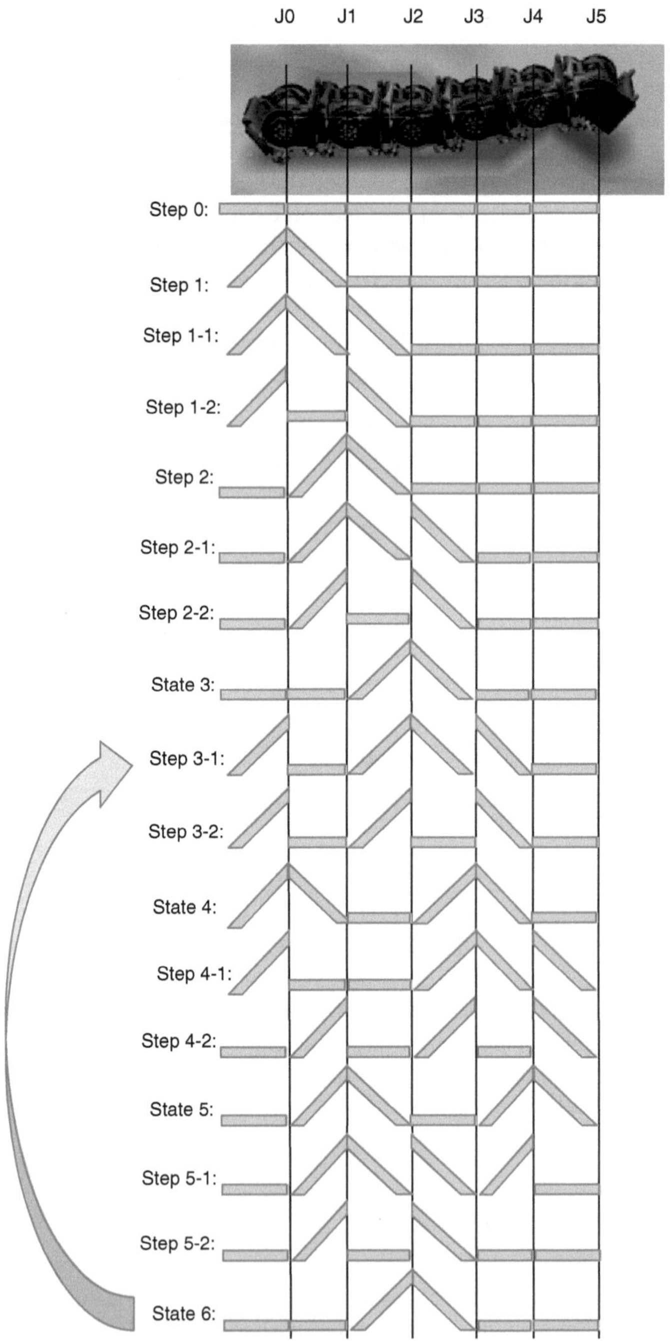

Fig. 5.8 Sinusoidal movement pattern of the robotic modules in organism 1 devised for the kinematic model. Variables $J0, \ldots, J5$ represent the movable joints between the robotic modules

Table 5.1 Experiment scenarios devised for organism 1

Case	Energy distribution	Mech. load distribution
C_1	$Even$	$Even$
C_2	$Uneven$	$Even$

ical load distribution among the robotic modules. It includes two scenarios: C_1 and C_2. Case C_1 simulated a scenario in which all the robotic modules of an organism were energetically equal or equally charged. In other words, the robotic modules were initialized with an even energy distribution and an even mechanical load distribution was used during their collective locomotion. To replicate a scenario in which the robotic modules autonomously assemble themselves in an organism with varying energetic status, case C_2 initialized the organism with an uneven energy distribution among the robotic modules. Due to the sinusoidal movement pattern of the robotic modules, case C_2 also used the even mechanical load distribution model during their collective locomotion.

For even and uneven initial energy distribution between the robotic modules, three different topologies of organism 1 were chosen. In this regard, Fig. 5.9a shows an even energy distribution between the robotic modules. Figure 5.9b, c show the randomly chosen uneven energy distributions between the robotic modules. Each box in Fig. 5.9 represents a robotic module and a numerical value inside represents the on-board battery state of charge. The legend bar shows the display pattern.

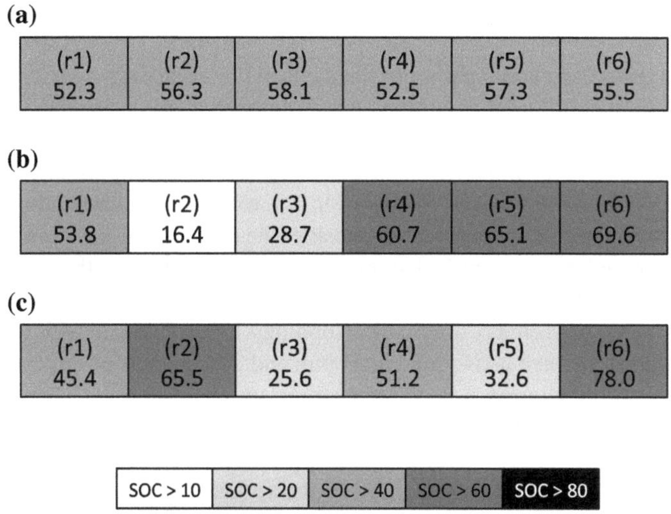

Fig. 5.9 Initial energy distribution among the REPLICATOR robotic modules in organism 1. **a** topology 1: even energy distribution. **b** topology 2: uneven energy distribution. **c** topology 3: uneven energy distribution

Figure 5.10 shows a screen-shot of the organism 1 simulation in the Replicator power flow simulator. The system dependent parameters in the simulation framework were adapted from the empirical results obtained from the real robotic modules. The parameters include *maximum battery pack charge capacity* of a robotic module that was chosen as 1,400 mAh, *fixed power consumption* as 200 mA at the nominal system voltage, and the variable power consumption values were obtained from the system power consumption model, described in Sect. 4.3.1. Using the system power consumption model, to lift the mechanical load of a robotic module a robotic module's hinge drive consumes 150 mA at the nominal voltage, for a load of two robotic modules it consumes 294 mA, with three robotic modules it consumes 806 mA, and to lift a load equivalent to the weight of four robotic modules its calculated current consumption was 1,575 mA.

On every time instant, the locomotion in the organism was simulated using the devised motion model matrix, MM_{Org1}. A motion step was incremented upon the successful completion of a motion cycle, a complete sinusoid. The organism's locomotion in the simulator was stopped as soon as the SOC of a robotic module in the organism reached 10 % of the maximum battery pack charge capacity, i.e.

$$\mathrm{SOC}_j(k) = 10\,\% \cdot \mathrm{SOC}_{\max}, \tag{5.8}$$

where the variable k represents the time instant and j represents the jth robotic module in the organism.

5.2.3 Discussion on Results

The effectiveness and applicability of the devised power sharing policies to achieve self-sufficiency in different conditions are evaluated on the basis of the simulation time and the covered motion steps by the organism. In this regard, Table 5.2 shows the simulation time of organism 1 in the three scenarios. Additionally, Table 5.3 shows the motion steps covered by organism 1 in each simulation run. In addition to these results, it is equally important to consider the current flow between the robotic modules of organism 1. Figure 5.11 shows the current flow through the energy sharing module of each robot with the three initial energy distribution topologies.

In the case C_1 with energy distribution topology 1, the organism with *static power sharing policy* remained active for 1 h, 51 min and 22 s and has covered 179 motion steps. With *power sharing policy* 1 the organism has covered 174 motion steps during 1 h, 48 min and 08 s. With these simulated values, now consider the current flow between the robotic modules shown in Fig. 5.11a. In the plots the measurements made with *static power sharing policy* are shown in blue, and with *power sharing policy* 1 are in red. With a single organism's power bus, i.e., *static power sharing policy*, the oscillations, i.e., the inward and outward current flow, at the energy sharing module show the robotic modules always shared their energy, even when all of them were energetically almost equal. In the beginning, robot r3 and r5 being slightly

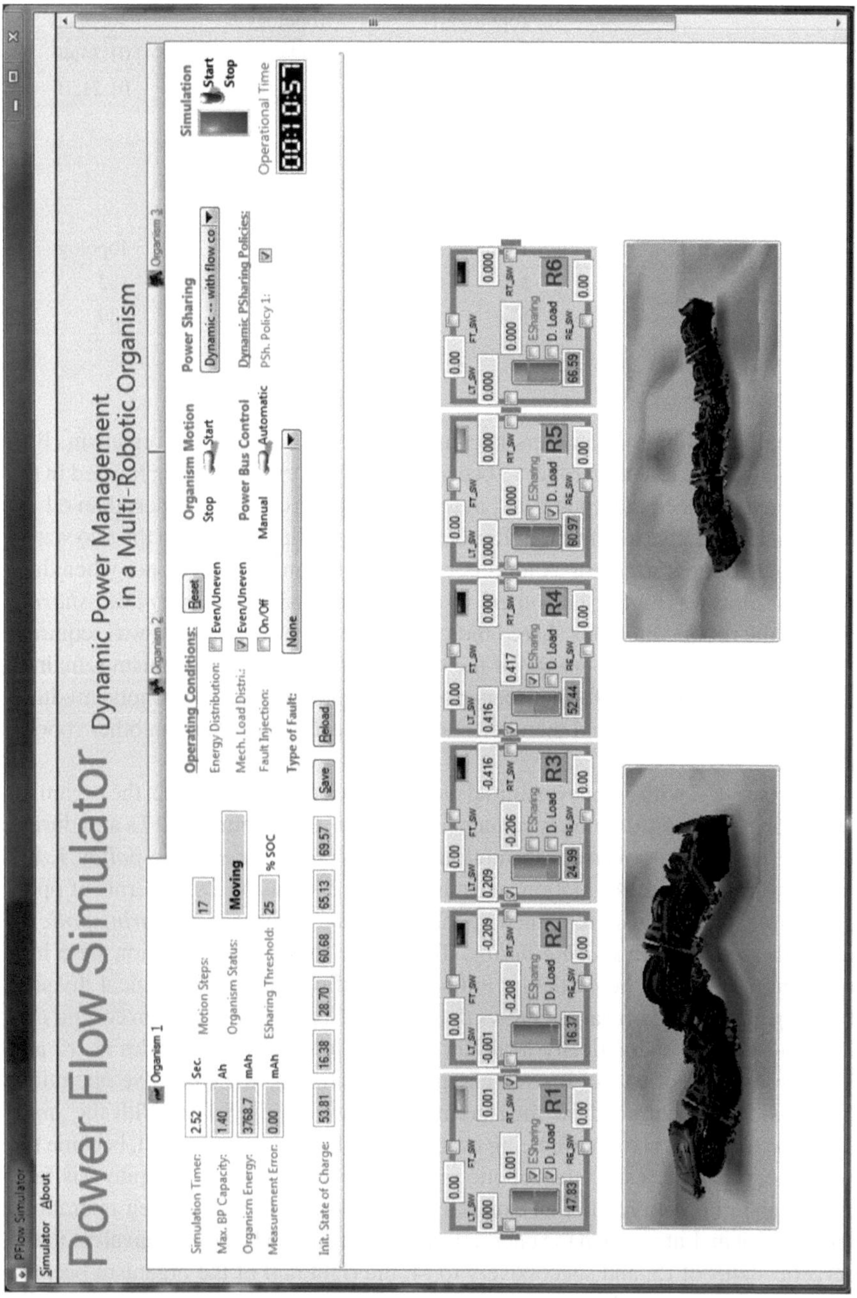

Fig. 5.10 A screen-shot of organism 1 simulation in the *Replicator Power Flow Simulator*

Table 5.2 Operational time of organism 1 during simulation in the case C_1 and C_2

Power sharing/distribution	Case C_1	Case C_2	
	Topology 1	Topology 2	Topology 3
Static	01:51:22	01:24:17	01:34:49
Policy 1	01:48:08	01:30:02	01:21:18

Table 5.3 Simulated motion steps covered by organism 1 in the case C_1 and C_2

Power sharing/distribution	Case C_1	Case C_2	
	Topology 1	Topology 2	Topology 3
Static	179	135	152
Policy 1	174	145	130

energetically healthier than others became the energy donor in the organism. But, approximately after 10 min, robot r3 became energy acceptor, as being located in the center, it experienced more load, i.e., in terms of lifting robotic modules, than others during their collective locomotion. Robot r2 and r5 oscillated between the two states, i.e., energy donor or acceptor. And, robot r1 and r6 became energy donor when their energetic status was higher than others. In comparison, using dynamic *power sharing policy* 1, the robotic modules shared their energy reserve only when it was required. By doing so, the on-board energy of the robotic modules in the organism remained preserved, whereas, in the former scenario, the energetically healthy robotic modules quickly lost their energy and then became energetically dependent on other robotic modules.

In the case C_2, with an uneven initial energy distribution topology 2, the organism with *static power sharing policy* remained active for 1 h, 24 min and 17 s and during this time period has covered 135 motion steps. With *power sharing policy* 1, the organism has covered 145 motion steps during 1 h, 30 min and 2 s. In terms of operational time, the organism remained operational longer with *power sharing policy* 1 than with *static power sharing policy*. To analyze the system behavior, here it is important to consider the particular initial energy distribution. The logical division of the organism into two halves shows that the right half of the organism comprising r4, r5 and r6, was energetically healthier than the left half that includes r1, r2 and r3. To observe the effects of uneven energy distribution on the collective operation, now consider the current flow measurements shown in Fig. 5.11b. With the *static power sharing policy*, robot r6, being energetically healthiest among all, became the energy donor to the rest of the robotic modules in the organism. Despite this fact, robot r5 and r4 do not require an external power source to continue their operations in the organism. Later, when the energetic status of robot r6 became equivalent to the energetic status of r5, and successively to r4, the right half of the organism became energy donor to the left half. In comparison, with *power sharing policy* 1, from the beginning, due to the weaker energetic status of robot r2, two sub-power buses were

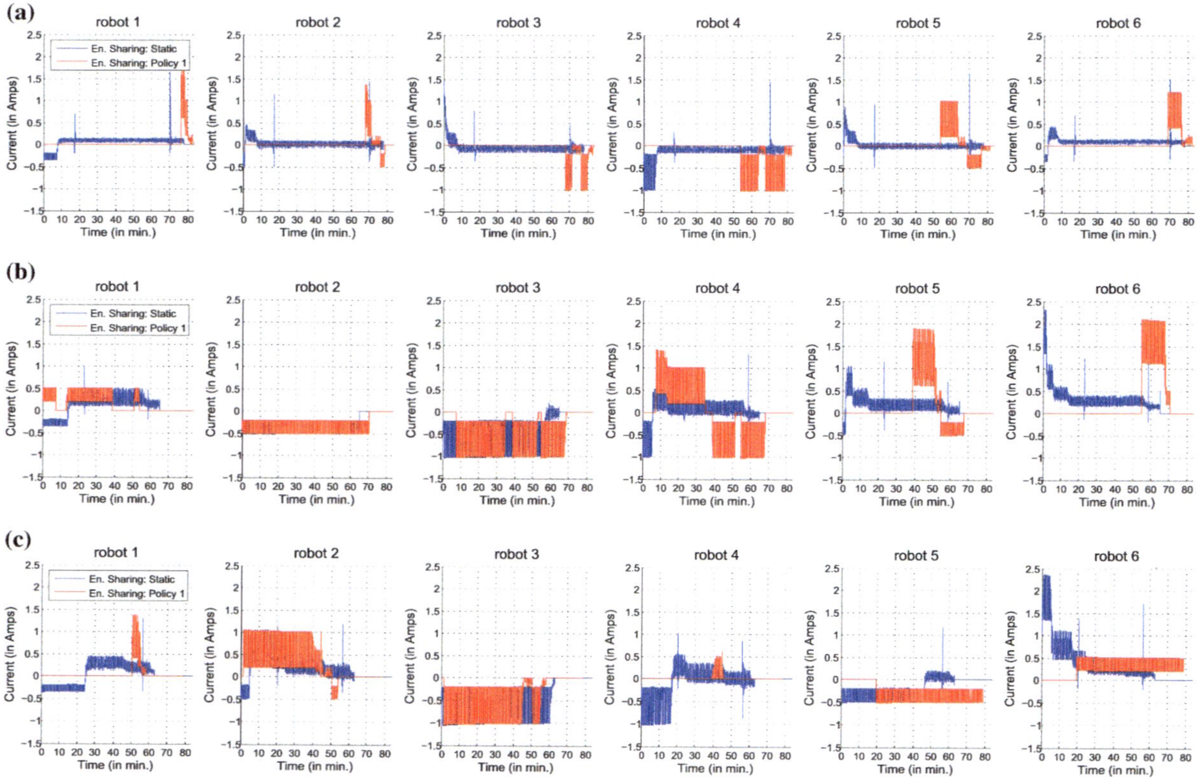

Fig. 5.11 Current flow through energy sharing module of each robot in organism 1 in different scenarios. The positive values in the plots show the outward current flow from the system and vice versa **a** case C_1 with initial energy distribution topology 1, **b** case C_2 with initial energy distribution topology 2, **c** case C_2 with initial energy distribution topology 3

established in the left half of the organism, i.e., a uni-directional power bus between robot r1 and r2 and a uni-directional power bus between robot r3 and r2. Since at robot r2 the source voltage of robot r1 was greater than the source voltage of r3, the power mainly flowed from robot r1 to r2. Later, when the energetic status of robot r3 dropped below the energy sharing threshold, another uni-directional sub-power bus was established between r4 and r3. And likewise, two uni-directional power buses were established, one at 39th min of the simulation between robot r5 and r4, and the second at 55th min between robot r6 and r5. This way the distributed energy in the organism was effectively utilized in a step-by-step manner, that allowed it to survive longer than in the earlier scenario.

In the case C_2, with uneven initial energy distribution topology 3, the organism with *static power sharing policy* has covered 152 motion steps in 1 h, 34 min and 49 s. With *power sharing policy* 1, it has covered only 130 motion steps in 1 h, 21 min and 18 s. In this form of uneven initial energy distribution topology, among the six robotic modules, three of them were energetically healthier than the rest, i.e., r2, r4, and r6. Noticeably, in this topology the energy was not concentrated in either half of the organism. Figure 5.11c shows the current flow measurements between the robotic modules with uneven energy distribution topology 3. With *static power sharing policy*, robot r6 being energetically healthiest became the energy donor to the rest of the robotic modules and due to this, it quickly lost its on-board energy reserve. The residual charge of the robotic module at the end of the simulation run showed with *power sharing policy* 1 the on-board charge of robot r6 was not fully utilized. It was because the energy sharing in the organism was based on the local energy distribution graph, that only allows power buses between direct neighbors. In such a scenario, it is expected that the organism using power sharing policy 2 will be able to cover same number of motion steps as with static power sharing policy.

The distribution of power between the robotic module using power sharing policy 1 showed promising results, in terms of operational time, especially in the scenario when one logical half of the organism was energetically healthier than the other. In the case when the energy distribution in the two logical halves of the organism was almost identical—uneven energy distribution topology 3—the simulation results show how the energy in an organism can be distributed during an organism formation and shared while keeping the individual energetic autonomy of the robotic modules.

5.3 Simulating Power Sharing in Organism 2

In the above, the proposed controlled power sharing mechanisms in a simple organism structure showed promising results. Now, to further explore their applicability, especially the *power sharing policy* 2 with different initial energy distribution topologies and an uneven mechanical load distribution, a relatively complex organism structure was chosen. The REPLICATOR modular robotic organism of type 2 comprises twelve Kabot modules, that are docked to each other in the form of a car—four limbs (actuator arms) and a central backbone—as shown in Fig. 5.12. The chosen organism's morphology is complex in the respect that the robotic modules in the

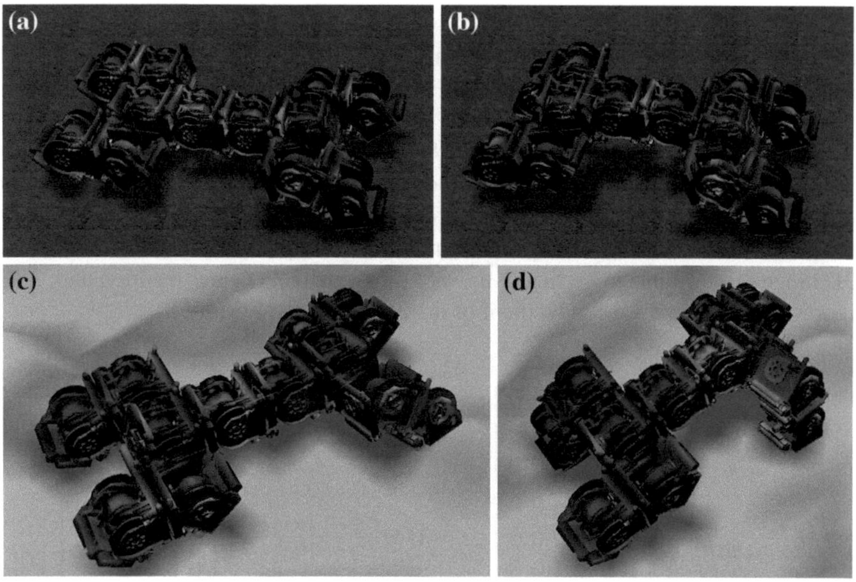

Fig. 5.12 REPLICATOR modular robotic organism of type 2 comprising twelve Kabot modules, **a** locomotion on a flat surface, **b** locomotion from another angle, **c** locomotion on a rough, uneven, terrain, **d** locomotion from another angle

four actuator arms during locomotion in the arena require more electrical power than the robotic modules docked in the central backbone. In Fig. 5.12a, b the organism is shown while moving on a flat surface. Similarly, Fig. 5.12c, d show the organism's locomotion on a rough, uneven, terrain.

5.3.1 Kinematic Model

The increased structural complexity of organism 2 resulted in more degrees of freedom and made it very difficult to mathematically model even a simple coordinated movement pattern. Since it was not in the scope of this work, the complex kinematics of the resultant system was ignored and a simple hypothetical kinematic model was devised. The kinematic model defined the movement pattern of the robotic modules by assigning a pre-calculated electronic load to each robotic module during their collective locomotion. During simulation, based on the active power sharing policy in the organism the assigned electronic load was then subtracted either from the host robot's battery pack or from the donor robot's battery pack.

The devised kinematic model for organism 2 consists of 12 motion patterns. Each motion pattern turn by turn actuates the robotic modules in the organism to achieve a desired movement. Figure 5.13 shows the step-wise movement patterns of the robotic modules in organism 2 during their collective locomotion. The organism movement patterns that try to replicate the caterpillar like movement consists of two phases:

erection and *relaxation*. From step 1 till step 6, the robotic modules in the organism collectively and individually erect themselves to attain a standing position. From step 7, the robotic modules step-by-step start restoring themselves to their nominal position (relaxing) until they complete a motion cycle in step 12.

Motion matrices MM_{Org2_Step1} till MM_{Org2_Step12} stepwise model the synchronous movement pattern of the robotic modules in organism 2. Here, "0" in the motion matrices represents *no motion*, "ϕ" represents the absence of a robotic module on the particular position in the organism structure, "1" represents the movement of the 3D hinge drive (joint) in vertically upward direction from center position and "-1" represents the downward movement of the 3D hinge drive. Motion

$$MM_{Org2\_Step1} = \begin{bmatrix} 1\,0\,\phi\,\phi\,0\,0 \\ \phi\,0\,0\,0\,0\,\phi \\ 1\,0\,\phi\,\phi\,0\,0 \end{bmatrix}, \quad MM_{Org2\_Step2} = \begin{bmatrix} 0\,1\,\phi\,\phi\,0\,0 \\ \phi\,0\,0\,0\,0\,\phi \\ 0\,1\,\phi\,\phi\,0\,0 \end{bmatrix},$$

$$MM_{Org2\_Step3} = \begin{bmatrix} 0\,0\,\phi\,\phi\,0\,0 \\ \phi\,0\,1\,0\,0\,\phi \\ 0\,0\,\phi\,\phi\,0\,0 \end{bmatrix}, \quad MM_{Org2\_Step4} = \begin{bmatrix} 0\,0\,\phi\,\phi\,0\,0 \\ \phi\,0\,0\,1\,0\,\phi \\ 0\,0\,\phi\,\phi\,0\,0 \end{bmatrix},$$

$$MM_{Org2\_Step5} = \begin{bmatrix} 0\,0\,\phi\,\phi\,1\,0 \\ \phi\,0\,0\,0\,0\,\phi \\ 0\,0\,\phi\,\phi\,1\,0 \end{bmatrix}, \quad MM_{Org2\_Step6} = \begin{bmatrix} 0\,0\,\phi\,\phi\,0\,1 \\ \phi\,0\,0\,0\,0\,\phi \\ 0\,0\,\phi\,\phi\,0\,1 \end{bmatrix}$$

matrices MM_{Org2_Step1}–MM_{Org2_Step6} model the movement pattern of the robotic modules in organism 2 during the *erection* phase, whereas motion matrices MM_{Org2_Step7} tillMM_{Org2_Step12} model the movement pattern of the robotic modules in organism 2 during the *relaxation* phase.

$$MM_{Org2\_Step7} = \begin{bmatrix} -1\,0\,\phi\,\phi\,0\,0 \\ \phi\;\,0\,0\,0\,0\,\phi \\ -1\,0\,\phi\,\phi\,0\,0 \end{bmatrix}, \quad MM_{Org2\_Step8} = \begin{bmatrix} 0\;-1\,\phi\,\phi\,0\,0 \\ \phi\;\;0\;\;0\,0\,0\,\phi \\ 0\;-1\,\phi\,\phi\,0\,0 \end{bmatrix},$$

$$MM_{Org2\_Step9} = \begin{bmatrix} 0\,0\;\phi\;\phi\,0\,0 \\ \phi\,0\,-1\,0\,0\,\phi \\ 0\,0\;\phi\;\phi\,0\,0 \end{bmatrix}, \quad MM_{Org2\_Step10} = \begin{bmatrix} 0\,0\,\phi\;\phi\;0\,0 \\ \phi\,0\,0\,-1\,0\,\phi \\ 0\,0\,\phi\;\phi\;0\,0 \end{bmatrix},$$

$$MM_{Org2\_Step11} = \begin{bmatrix} 0\,0\,\phi\,\phi\,-1\,0 \\ \phi\,0\,0\,0\;\;0\;\,\phi \\ 0\,0\,\phi\,\phi\,-1\,0 \end{bmatrix}, \quad MM_{Org2\_Step12} = \begin{bmatrix} 0\,0\,\phi\,\phi\,0\,-1 \\ \phi\,0\,0\,0\,0\;\;\phi \\ 0\,0\,\phi\,\phi\,0\,-1 \end{bmatrix}$$

The effects of initial energy distribution and dynamic power sharing on the organism locomotion on different surfaces were explored using the corresponding load matrices attached to the motion model devised for the two surface, *flat* and *rough terrain*. The locomotion on a flat surface was an attempt to replicate a scenario in which all the actuator robotic modules apply an equal amount of force during their collective locomotion. In this regard, an even distribution of mechanical load on

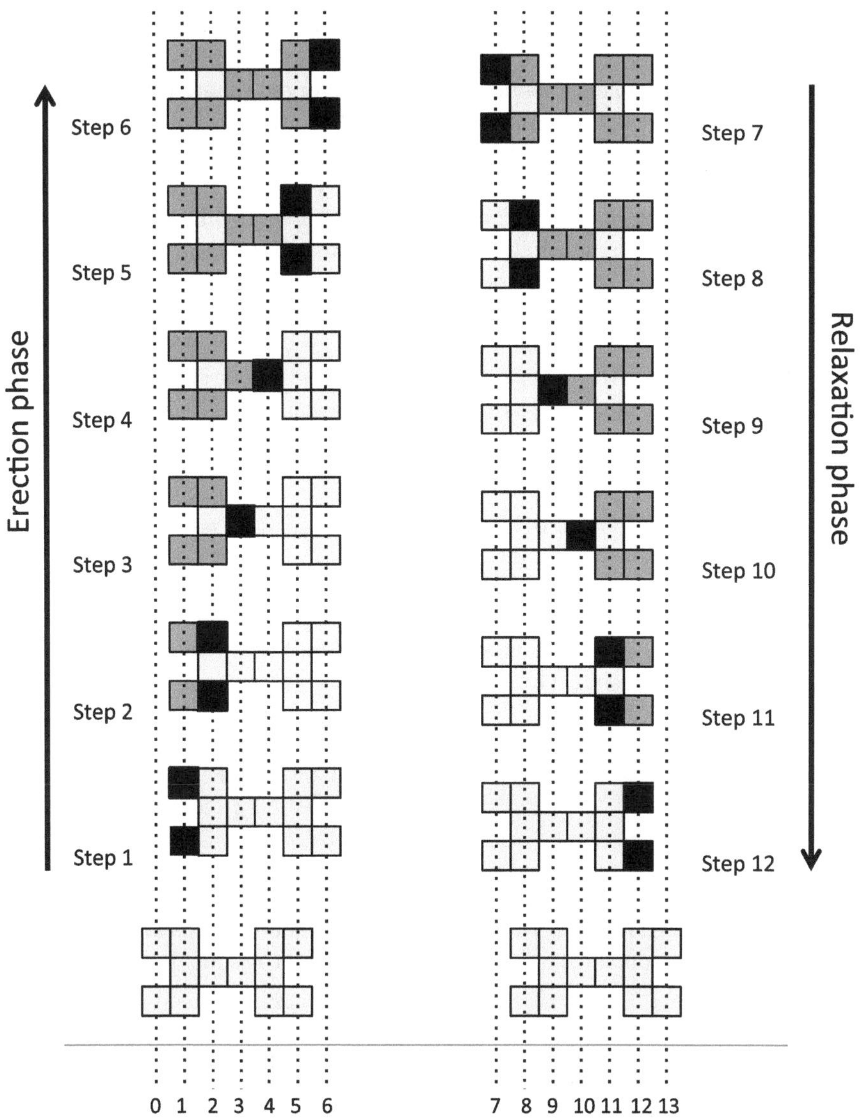

Fig. 5.13 Stepwise movement pattern of the robotic modules in organism 2—from *left* to *right*—to model their collective locomotion in the arena. The *horizontal* axis shows the discrete time steps representing a complete motion cycle. The organism movement patterns that try to replicate the caterpillar like movement consists of two phases: *erection* and *relaxation*. For a better illustration, the collective movements of the robotic modules are shown with the following colors: *white, black,* and *grey*. The *white color* shows their nominal position with respect to the 3D motion. The *black color* of a robotic module depict its 3D motion at the particular motion step, and the *grey color* depict their erected state in the organism. From step 1 till step 6, the robotic modules in the organism collectively and individually erect themselves to attain a standing position. From step 7 till step 12, again *black color* of the robotic modules depicts their move in restoring their nominal position (relaxing), until they complete a motion cycle

the actuator robotic modules was applied to simulate the organism locomotion. In contract, the locomotion on a rough terrain was modeled to replicate the scenario in which some of the actuator robotic modules in the organism are required to apply more force than the rest. Consequently, it applied an uneven load distribution among the four actuator arms of organism 2.

Load matrix $LM_{Org2\_Even\_erect}$, which associated with the movement patterns in the erection phase, defines an even distribution of mechanical load among the four actuator arms of organism 2. Similarly, load matrix LM_{Org2_relax} defines the associated mechanical load distribution between the robotic modules during the relaxation phase. In the simulation, the devised motion model used the load matrix $LM_{Org2\_Even\_erect}$ during the motion steps MM_{Org2_Step1} till MM_{Org2_Step6}, and likewise, load matrix LM_{Org2_relax} was used during the motion steps MM_{Org2_Step7} to MM_{Org2_Step12} to obtain the associated load at each robotic module.

$$LM_{Org2\_Even\_erect} = \begin{bmatrix} 3\,2\,0\,0\,2\,3 \\ 0\,0\,3\,3\,0\,0 \\ 3\,2\,0\,0\,2\,3 \end{bmatrix}, \qquad LM_{Org2\_relax} = \begin{bmatrix} 1\,1\,0\,0\,1\,1 \\ 0\,0\,1\,1\,0\,0 \\ 1\,1\,0\,0\,1\,1 \end{bmatrix}$$

Similarly, for locomotion on a rough terrain—with crust and troughs—load matrix $LM_{Org2\_Uneven\_erect}$ defines an uneven mechanical load distribution among the four actuator arms of organism 2, i.e.,

$$LM_{Org2\_Uneven\_erect} = \begin{bmatrix} 3\,2\,0\,0\,3\,4 \\ 0\,0\,3\,4\,0\,0 \\ 4\,3\,0\,0\,2\,3 \end{bmatrix}.$$

In the simulation experiments, load matrix $LM_{Org2\_Uneven\_erect}$ was used during the motion steps MM_{Org2_Step1} till MM_{Org2_Step6}, and during the *relaxation phase*, load matrix LM_{Org2_relax}, was used during the motion steps MM_{Org2_Step7}–MM_{Org2_Step12} to obtain the associated load at each robotic individual in the organism.

5.3.2 Simulation Setup

The simulation setup for organism 2 used the initial energy distribution and the mechanical load associated with the locomotion of the robotic modules according to the test cases in Table 5.4 to analyze the effectiveness of the devised power sharing policies.

In the simulation experiments the robotic modules in organism 2 were initialized with an even and two uneven energy distribution topologies. Figure 5.14 shows the chosen initial energy distribution topologies for organism 2. Each box represents a robotic module and a numerical value inside is representing the on-board battery state of charge. The even initial energy distribution shown in Fig. 5.14a, was chosen to simulate a scenario in which all the robotic modules are energetically almost equal or equally charged. On the contrary, the two randomly chosen uneven initial energy

Table 5.4 Experiment scenarios devised for organism 2

Case	Energy distribution	Mechanical load distribution
C_1	Even	Even
C_2	Even	Uneven
C_3	Uneven	Even
C_4	Uneven	Uneven

distribution topologies, as shown in Fig. 5.14b, c, were used to simulate the scenarios in which the robotic modules with varying energetic status autonomously assembled or re-assembled themselves in the particular morphology.

Figure 5.15 shows a screen-shot of organism 2 in the *Replicator Power Flow Simulator*. The simulation front-end provides an access to monitor and control the different system parameters, e.g., selection of initial energy distribution, mechanical load distribution, power sharing policy between the robotic modules, current flow through docking interfaces and energy sharing module of each robot, etc.

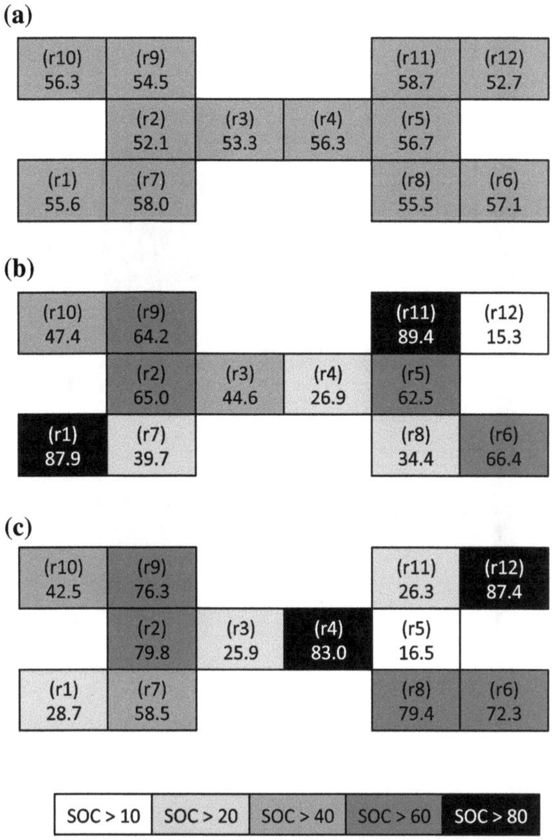

Fig. 5.14 Initial energy distribution topologies among the robotic modules of organism 2. The legend bar shows the used color scheme. **a** topology 1: even energy distribution among the robotic modules. **b** topology 2: uneven energy distribution among the robotic modules. **c** topology 3: uneven energy distribution among the robotic modules

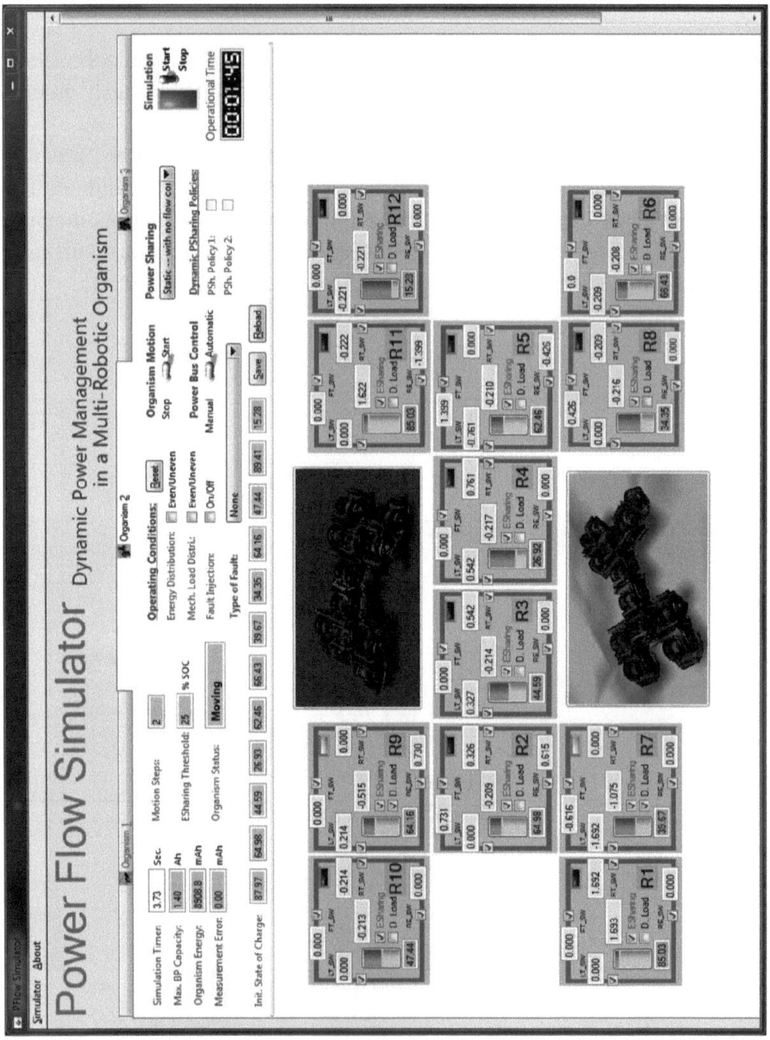

Fig. 5.15 A screen-shot of organism 2 simulation in the *replicator power flow simulator*

5.3.3 Discussion on Results

During each simulation run, the current flow values at the energy sharing module and the four docking interfaces of each robotic module in the organism were logged on every iteration. From the simulation results, Tables 5.5 and 5.6 show the operational time and the motion steps covered by organism 2, respectively.

In the case C_1, the organism with an even initial energy and an even mechanical load distribution, using *static power sharing policy* has covered 182 motion steps in 2 h, 30 min and 8 s. Using *power sharing policy* 1, it covered 175 motion steps in 2 h, 24 min and 48 s, and with *power sharing policy* 2, it was able to cover 177 motion steps in 2 h, 25 min and 6 s. Now, to explore the effects of power sharing on the obtained results consider the current flow measurements shown in Fig. 5.16. The positive current values in the plots show the outward current flow from a particular module and vice versa. With *static power sharing policy*, a single omni-directional power bus allowed the robotic modules to share their on-board battery charge with each other. The discharging of multiple battery packs in parallel irrespective of their locality in the organism therefore created oscillations at the energy sharing module of each robot. These oscillations, i.e., inward and outward current flow through a system, visible in the current flow measurements shown in Fig. 5.16 depict the relative energetic role of a robotic module—either as energy donor or acceptor—in the organism. Whereas, using the dynamic *power sharing policies*, sub-power buses were established between the robotic modules only at different simulation instances when power sharing was required. By doing so, the on-board battery charge of the robotic modules remained preserved which in turn empowers them to keep their energetic autonomy in case of a sudden breakdown in the organism.

In the case C_2, the locomotion in organism 2 was simulated using an even initial energy distribution and an uneven mechanical load distribution among the robotic

Table 5.5 Operational time of organism 2 during simulation runs

Energy sharing/	Case C_1	Case C_2	Case C_3		Case C_4	
distribution	Topology 1	Topology 1	Topology 2	Topology 3	Topology 2	Topology 3
Static	2:30:08	2:13:09	01:54:58	2:16:12	1:43:04	2:07:21
Policy 1	2:24:48	2:09:39	01:50:01	2:21:53	1:35:01	2:04:08
Policy 2	2:25:06	2:08:48	01:51:37	2:22:01	1:41:57	2:03:47

Table 5.6 Covered motion steps by organism 2 during simulation runs

Energy sharing/	Case C_1	Case C_2	Case C_3		Case C_4	
distribution	Topology 1	Topology 1	Topology 2	Topology 3	Topology 2	Topology 3
Static	182	161	140	165	125	154
Policy 1	175	156	135	171	115	150
Policy 2	177	155	136	172	124	149

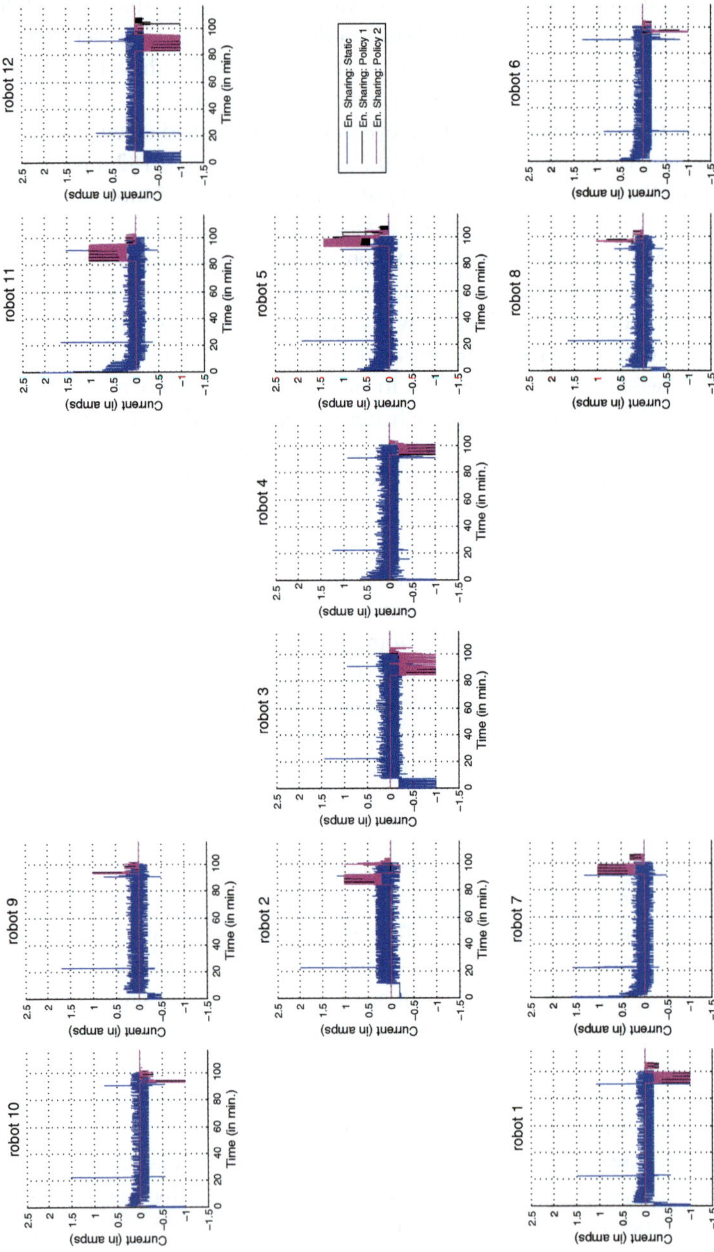

Fig. 5.16 Current flow through the energy sharing module of each robot in the case C_1—with initial energy distribution topology 1

modules. The organism 2 using *static power sharing policy* has covered 161 motion steps in 2 h, 13 min and 09 s. With *power sharing policy* 1, it covered 156 motion steps in 2 h, 09 min and 39 s, and with *power sharing policy* 2, it was able to cover 155 motion steps in 2 h, 08 min and 48 s. A direct effect of an uneven mechanical load distribution can be seen in the form of fewer motion steps in the three scenarios in comparison with the case C_1. With *static power sharing policy*, again, a single omni-directional organism's power bus connected the on-board battery pack of the robotic modules in parallel. In such a configuration, firstly, the battery with higher SOC was drained as long as its SOC remained higher than the rest in the organism. Figure 5.17 shows the current flow at the energy sharing module of each robotic module in organism 2. With the two dynamic power sharing policies, the current flow between the robotic modules remained localized, i.e., between the direct neighbors. This is the reason, why the robotic modules shared their battery charge even longer than it was with the *static power sharing policy*.

Now, consider the collective system behavior in the case C_3—with an uneven energy distribution topology 2 and an even mechanical load distribution. The organism using *static power sharing policy* has covered 140 motion steps in 1 h, 54 min and 58 s. With *power sharing policy* 1, it covered 135 motion steps in 1 h, 50 min and 1 s, and lastly, with *power sharing policy* 2, it was able to cover 136 motion steps in 1 h, 51 min and 37 s. Figure 5.18 shows the current flow measurements between the robotic modules of organism 2 in the case C_3. For a detailed analysis of the collective system behavior during the simulation runs, it is important to consider the initial energy distribution in the organism, as it defined the current flow between the robotic modules. In this regard, consider the two logical halves of organism 2—between robot r3 and robot r4. The logical splitting of organism showed that the ratio of energetically healthy and weaker robotic modules in term of their initial SOC in the two halves was almost identical. With this observation, now consider the current flow between the robotic modules with *static power sharing* policy, because of a single organism's power bus in the first 15 min of the simulation, robot r1 and robot r11 donated their battery charge to the rest of the robotic modules in the organism. Later in the simulation, other robotic modules shared their on-board energy reserve in the organism when the energetic status of robot r1 and robot r11 dropped to their energetic level—battery SOC. In comparison, with the two *dynamic power sharing policies*, the on-board energy of the robotic modules was consumed locally. From the residual charge of the robotic modules at the end of the simulation runs, it was observed that the on-board energy reserve of robot r1 with the two *dynamic power sharing policies* was not completely utilized as in the earlier scenario. Because of this, the organism has covered slightly fewer number of motion steps than in the earlier scenario.

Now consider an uneven initial energy distribution topology in which the two logical halves of the organism were energetically not identical. In this regard, the robotic modules in organism 2 were initialized with the values proposed in energy distribution topology 3. In the particular energy distribution topology the right half of the organism was energetically healthier than the left half. The organism using *static power sharing policy* has covered 165 motion steps in 2 h, 16 min and 12 s.

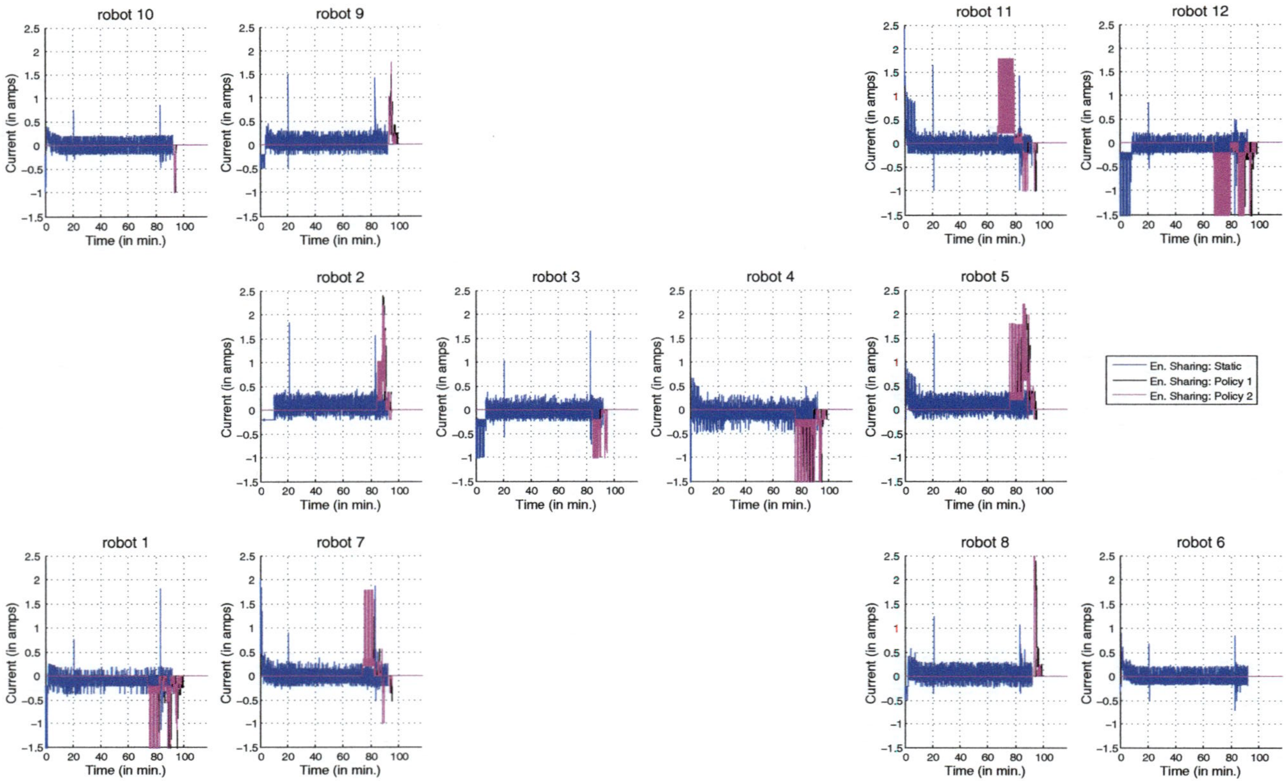

Fig. 5.17 Current flow through the energy sharing module of each robot in the case C_2—with initial energy distribution topology 1

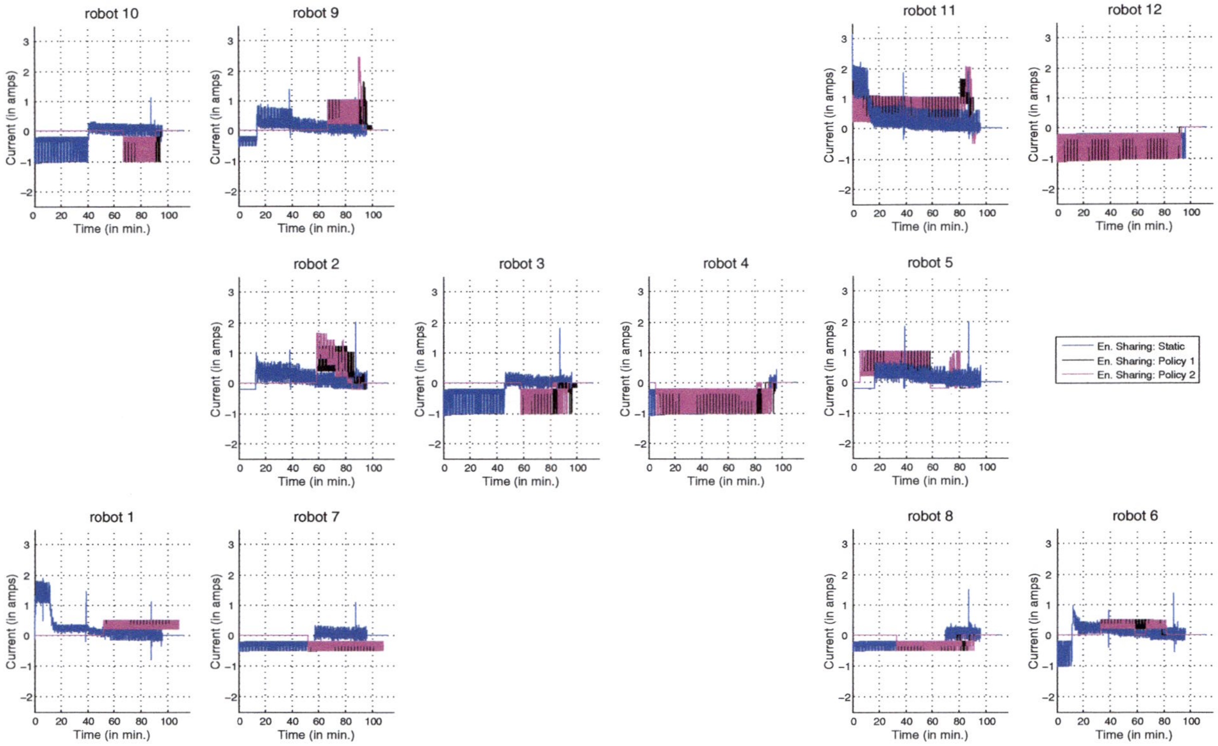

Fig. 5.18 Current flow through the energy sharing module of each robot in the case C_3—with initial energy distribution topology 2

With *power sharing policy* 1, it has covered 171 motion steps in 2 h, 21 min and 53 seconds. And, using *power sharing policy* 2, it has covered 172 motion steps in 2 h, 22 min and 1 s. Figure 5.19 shows the current flow between the robotic modules of organism 2 in the case C_3. Notably, using *static power sharing* policy because of a single organism's power bus, in the beginning, robot r9 and r6 consumed power from the organism's power bus even though they carry enough on-board battery charge. On the other side, using *power sharing policies* 1 and 2, robot r6 remained neutral, i.e., neither consumed nor donated its battery charge, for a longer period of time in the simulation, whereas, robot r9 donated energy in the second half of the simulation. From the residual charge of the robotic modules at the end of the simulation, it was observed that using *static power sharing* the energetically weakest robot, i.e., robot r5, stopped the organism locomotion in the arena. But, with the two *dynamic power sharing policies* the localized power flow between the robotic modules allowed them to prolong their cooperation which resulted in more motion steps than with the static power sharing.

In the case C_4, to mimic the locomotion on a rough terrain, the locomotion in organism 2 was simulated using an uneven mechanical load distribution with the two uneven initial energy distribution topologies. Figure 5.20 shows the current flow between the robotic modules in the case C_4 using initial energy distribution topology 2. The organism using *static power sharing* policy has covered 125 motion steps in 1 h, 43 min and 4 s, using *power sharing policy* 1, covered 115 motion steps in 1 h, 35 min and 1 s, and lastly, using *power sharing policy* 2, covered 124 motion steps in 1 h, 41 min and 57 s. A direct effect of locomotion on a rough terrain—with uneven mechanical load distribution—can be seen in the form of fewer motion steps in comparison with the motion steps covered on a flat surface with the same initial energy distribution. In particular, with uneven mechanical load distribution the controlled power flow between the robotic modules —with *power sharing policy* 2—allowed the robotic modules to cover almost equivalent number of motion steps as with the *static power sharing policy*. In other words, this shows that in the particular scenario an organism may choose between the two controlled power sharing policies, depending upon the situation, to obtain a longer operational time.

In the last scenario, the locomotion of organism 2 was again simulated on a rough terrain, but the robotic modules were initialized with energy distribution topology 3. The organism 2 using *static power sharing policy* has covered 154 motion steps in 2 h, 7 min and 21 s. Using *power sharing policy* 1, it has covered 150 motion steps in 2 h, 4 min and 8 s, and using *power sharing policy* 2, it covered 149 motion steps in 2 h, 3 min and 47 s. Figure 5.21 shows the current flow measurements between the robotic modules of organism 2 in the case C_4. By comparing the results obtained in the case C_3—energy distribution topology 3 and an even mechanical load distribution—it is evident that with the uneven load distribution, the organism 2 was not able to cover an equivalent number of motion steps with the two controlled *power sharing policies*.

In general, the simulation results show that the energetic behavior of the robotic modules, in terms of number of covered motion steps and the operational time, with *static* and *dynamic* power sharing policies was almost identical under different initial energy distribution and mechanical load distribution scenarios. In comparison,

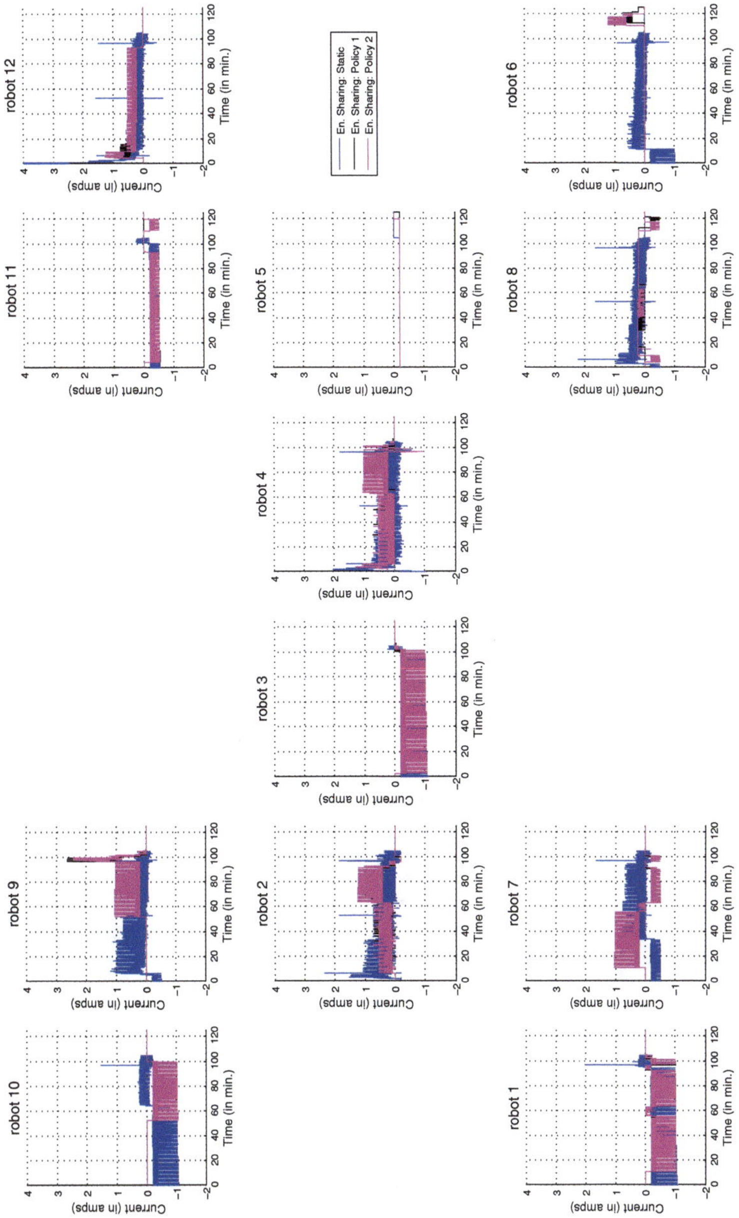

Fig. 5.19 Current flow through the energy sharing module of each robot in the case C_3—with initial energy distribution topology 3

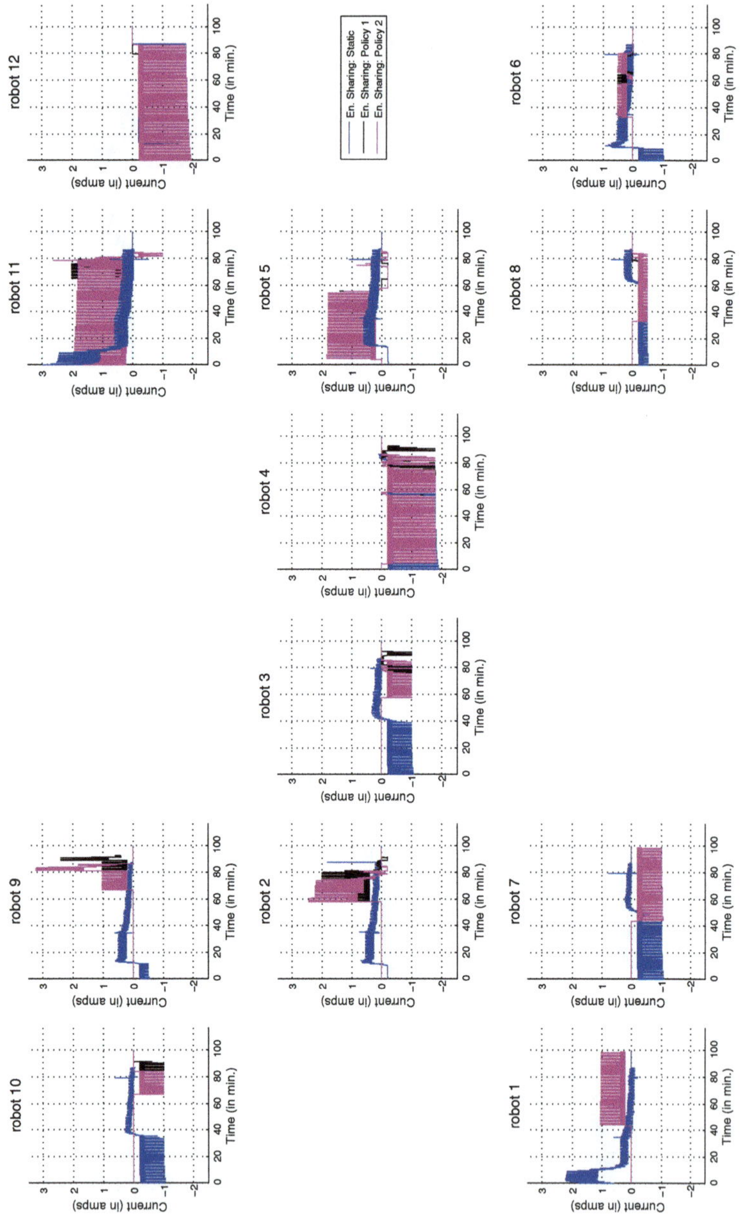

Fig. 5.20 Current flow through the energy sharing module of each robot in the case C_4—with initial energy distribution topology 2

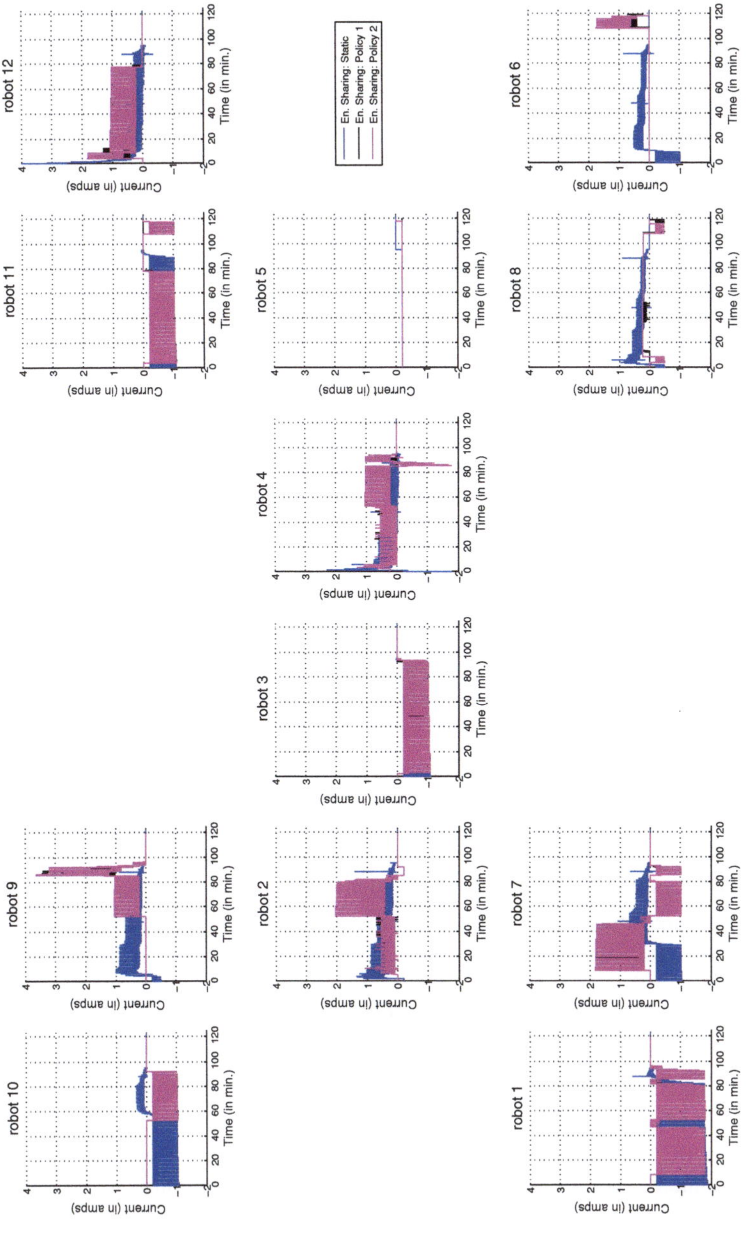

Fig. 5.21 Current flow through the energy sharing module of each robot in the case C_4—with energy distribution topology 3

the devised dynamic power sharing policies provide more flexibility in terms of energy sharing and therefore can be chosen on the fly by an autonomous modular robotic organism during different behavioral states, as identified in Sect. 3.2.2. For instance, in a partial breakdown behavioral state when static power sharing topology is not possible, an organism using dynamic power sharing topology can accomplish its objective without splitting or reorganizing the robotic modules, that in addition requires energy resources. In this regard, proactive power management policies can be developed at the application software layer for the selection of an appropriate power sharing mechanism among the robotic modules with the changing internal and external conditions. In fact, the simulation experiments helped to explore and understand the effects of energy distribution between the robotic modules of an organism. That is, how should multiple robotic modules with varying energetic status dock to each other during an organism formation in order to gain a longer operational time under different conditions. In other words, the effects of initial energy distribution on the collective operations of robotic modules in an organism.

5.3.4 Limitation of the Simulation Framework

The *Replicator Power Flow Simulator* in the limited frame still provided a platform to examine the power sharing behavior of modular robotic organisms that no other simulation tool offered at the time of writing this work. Aside other limitations, e.g., kinematic model, simulation of battery pack, fault injection and simulation, etc., on every iteration the SPICE simulator engine in the simulation framework requires to process the generated net-list from the beginning to obtain the nodal analysis of the electronic circuit that defines the interconnection of robotic modules in an organism. The time required for each iteration step during the simulation to obtain the nodal analysis in the case of organism 2 varied between 10 s to 5 min. As a consequence, the cumulative time required to process a complete simulation of organism 2—till the time the available energy can be used for collective locomotion—thus varied between 32 to roughly 50 h. Because of the limited resources and a considerable simulation time some of the planned experiments had to be dropped therefore.

5.4 Fault Tolerance Simulation Experiments

In the designed experiments, the effects of an abrupt system failure on the operations of multiple robotic modules was analyzed by simulating a short circuit condition at three different locations in the organism. And, by exploiting the inherent redundancy in the system, a sensor failure at a docking unit of a robotic module in the organism was located.

5.4.1 Experimental Setup

As mentioned in the beginning, from the perspective of self-sufficiency, the focus was laid only on those system components that in a faulty state can affect the power flow between the robotic modules of an organism. For this purpose, a SPICE simulation tool was used to simulate and explore the multiple effects of faults on dynamic power sharing in an organism. Figure 5.22 shows the block diagram of an organism with an omni-directional power bus between the robotic modules, chosen for the simulation experiments. The particular color scheme in the Fig. 5.22 depicts the energetic status of the robotic modules in the organism. That is, robot $r1$, $r2$, $r5$, and $r6$, shown in *white* color was considered as energetically weaker than the two robotic modules, $r3$ and $r4$, where robot $r4$ was supposed to be energetically healthier than the robot $r3$.

For the simulation of power flow and different kinds of faults between the robotic modules, parts of the power management components described in Sect. 4.2, was used in LTSpice simulation software from Linear Technology. Figure 5.23 shows the electronic components of the proposed power management system of a robotic module in LTSpice. It include two docking interfaces, an energy sharing module and a simple voltage source. Each docking unit (interface) is a combination of a diode and a current limiter. For experiment purposes, at each robotic module, a resistive load that consumed 250 mA of continuous current at the nominal system voltage was used as the fixed load and a resistive load that consumed 1 A of continuous current was used in place of actuators, as already described in Sect. 4.3.1.

For benchmark purposes, i.e., to compare and analyze the different effects of a fault on the operations of multiple robotic modules in an organism, the collective system behavior during power sharing in the absence of any fault or a component failure was at first simulated. In this regard, Fig. 5.24 shows the current flow at the energy sharing module of each robotic module in the organism. Positive values of the current show its outward flow from the system and vice versa. In the particular scenario, with an omni-directional power bus the robotic modules $r4$ and $r3$ donated their battery charge to the rest of the robotic modules in the organism. From the simulated current flow values, a continuous current of ≈4.7 A flown from robot $r4$ to its right and left docking interface and successively from robot $r3$ a continuous current of ≈900 mA was flowing to its left docking interface, i.e., to robotic modules $r2$ and $r1$.

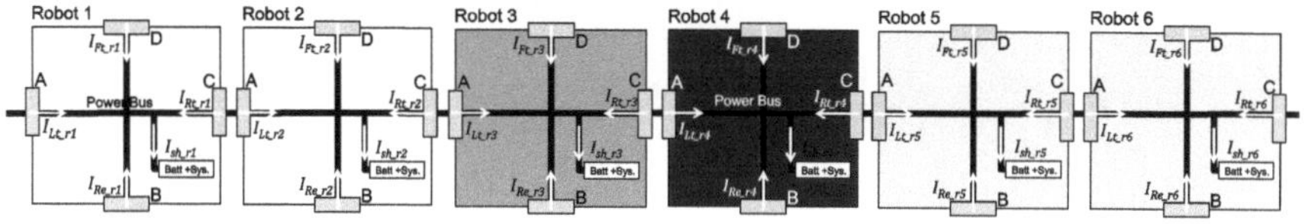

Fig. 5.22 An omni-directional organism's power bus between the robotic modules of an organism

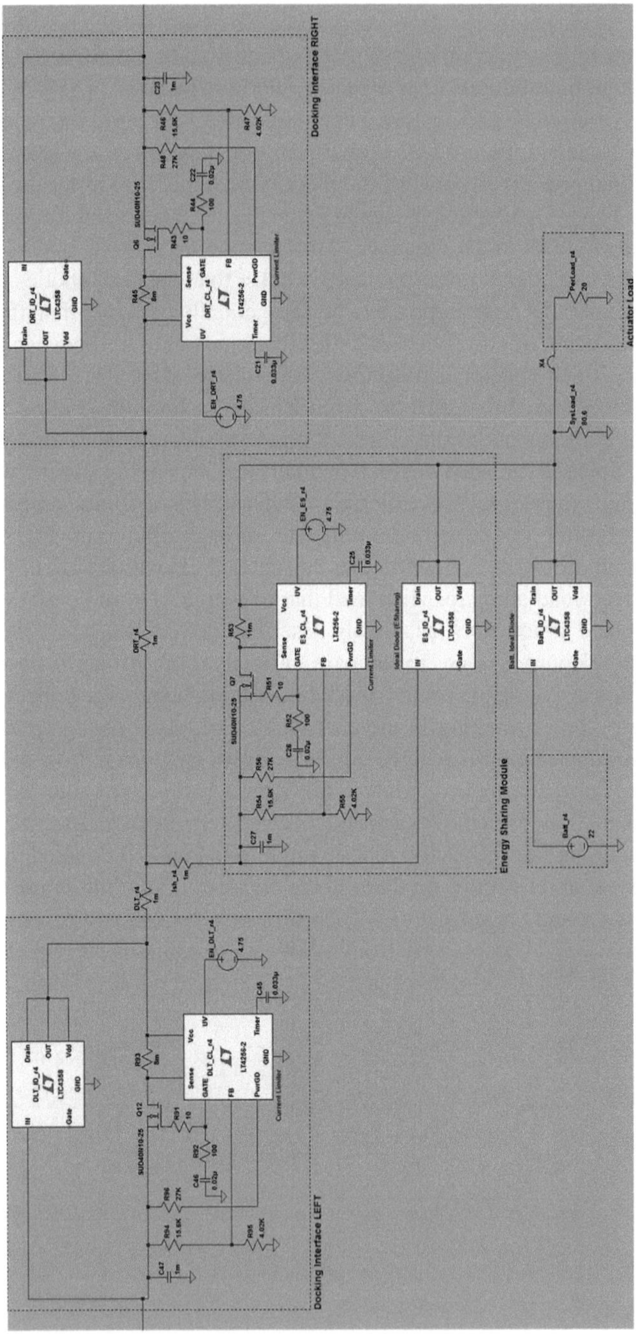

Fig. 5.23 Parts of the proposed power management system of a robotic module in LTSpice

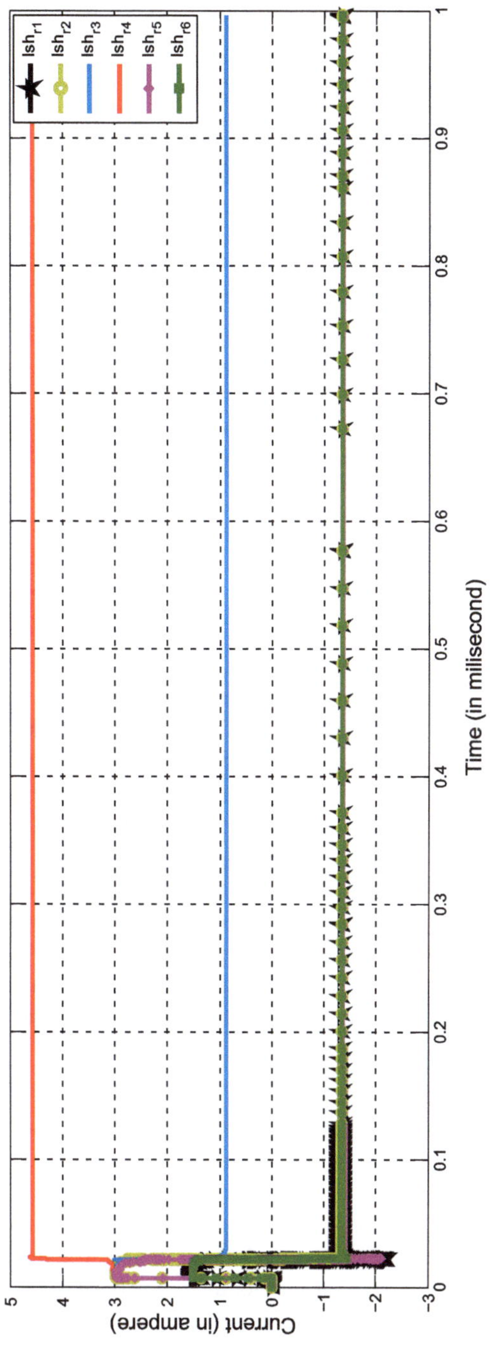

Fig. 5.24 Current flow through the energy sharing module of each robot in the absence of any fault

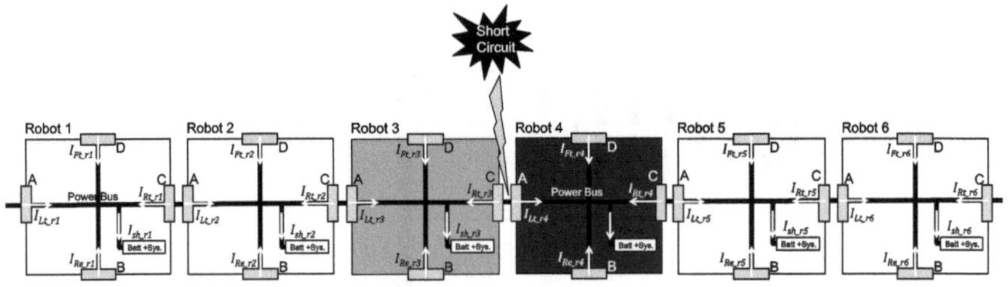

Fig. 5.25 Block diagram highlighting the injected fault location in the organism

5.4.2 Fault Occurrence Between the Docking Interfaces

In this scenario, it was tried to mimic an exogenous fault condition in which because of a faulty or an improper mechanical assembly, a short circuit appeared between the docking contacts of robot $r3$ and $r4$. The particular fault condition was replicated by adding a short circuit in the robot electronics to simulate its effects on the organism's behavior. Figure 5.25 shows the block diagram of organism, highlighting the particular fault location.

Fault Injection

After the fault injection in the robot electronics the SPICE simulation was performed to record the current flow between the robotic modules. Figure 5.26 shows the current flow measurements through the energy sharing module of each robot in the presence of a short circuit. Comparing the current flow values with the values made in the absence of any fault reveals the instant effect of a short circuit on the collective system behavior. That is, because of a short circuit the current flow through the energy sharing module of robot $r4$ was dropped to ≈ 2.8 A, since its left docking interface current limiter instantly disconnected the particular current flow path. The spikes visible in the current flow are because of the current limiter, as it tried to re-establish the path after the expiry of a timer, embedded in the hardware, but failed in its attempt. In response, the current flow through the energy sharing module of robot $r3$ increased, as it became the only energy donor to the robotic modules $r2$ and $r1$. In addition, the effects of the particular fault condition can also be seen in the form of spikes on the current flow through the energy sharing module of robot $r2$, which is not directly connected to the faulty component.

Fault Detection and Isolation

After adapting the procedures detailed in algorithm 1 and 2, the SPICE simulation was carried out with the new configurations. Figure 5.27 shows the current flow between the robotic modules after fault detection, identification and isolation. As a result of adapting the new configurations the right and left docking interface of robot $r3$ and $r4$ are turned OFF, respectively. In response, the spikes appeared in the current flow throughthe energy sharing modules of robot $r2$, $r3$ and $r4$ before the

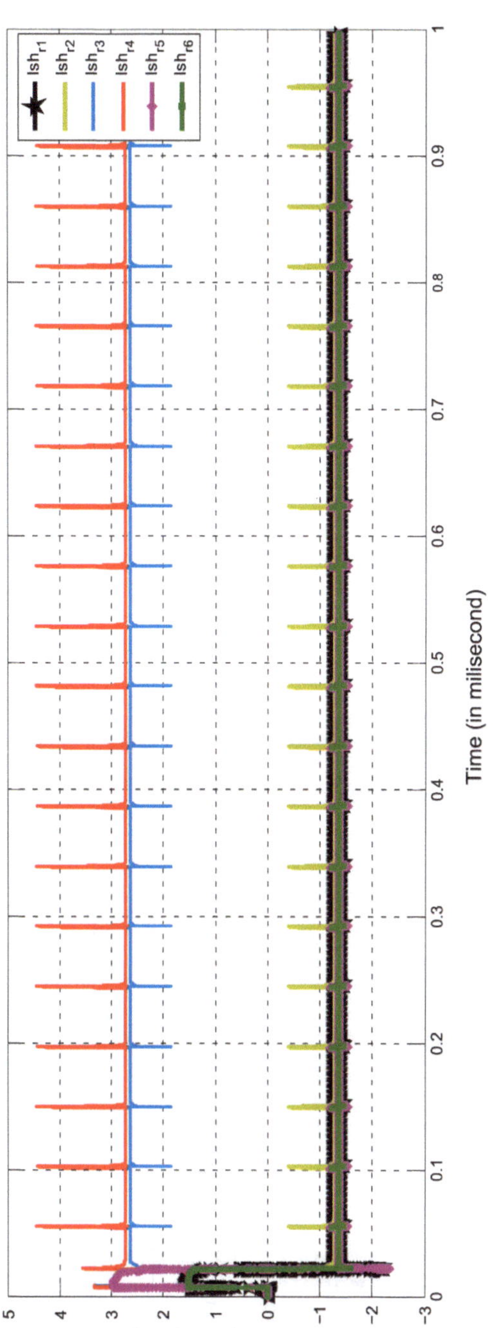

Fig. 5.26 Current flow through the energy sharing module of each robot in the presence of a short circuit between the docking interface of robot $r3$ and $r4$

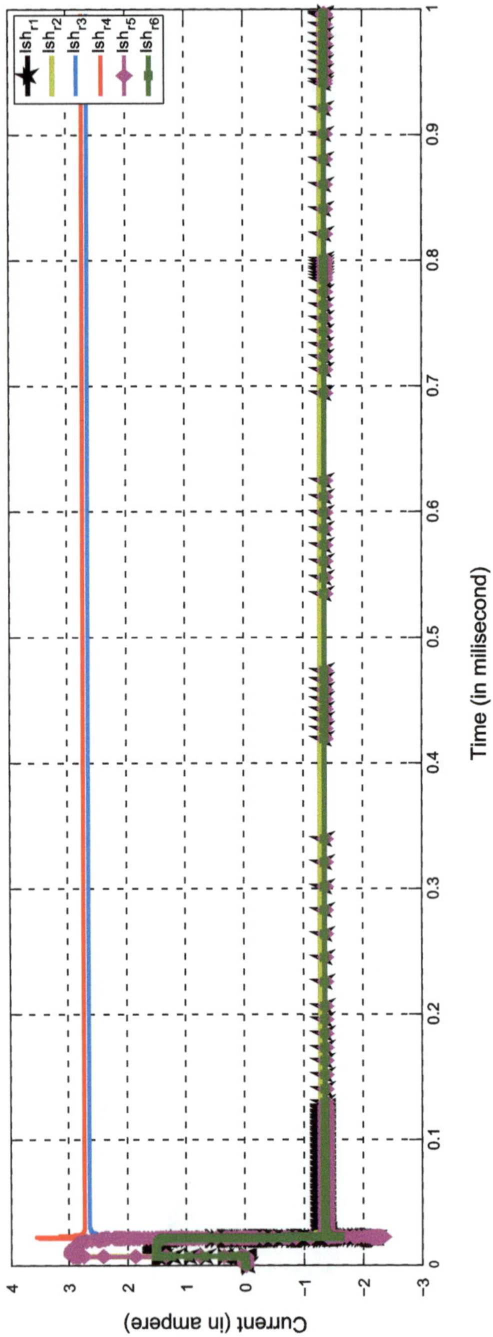

Fig. 5.27 Current flow between the robotic modules after the detection and isolation of an occurred fault

fault isolation are now no longer visible. This is due the fact that the current flow path through the respective docking interfaces of robot $r3$ and $r4$ was now disconnected.

5.4.3 Fault Occurrence Before a Docking Interface

The particular fault condition was created to determine the impact of an endogenous fault in which a short circuit appeared between the energy sharing module and one of the docking interface of a robotic module. Figure 5.28 is graphically highlights the fault location and the position of the faulty robotic module in the organism—between the energy sharing module and the right docking interface of robot $r3$.

Fault Injection

After injecting the particular fault in the robot electronics, the simulation was conducted using LTSpice to observe the collective system behavior. Figure 5.29 shows the current flow between the robotic modules after the introduction of a short circuit at the specified location. Analyzing the current flow data, i.e., before and after the occurrence of a fault, showed that because of a short circuit at robot $r3$, the current flow from robot $r4$ to the robotic modules docked at its left docking interface was completely stopped. In fact, the effects of the particular fault can be observed in the form of spikes in the current flow data obtained at the energy sharing module of robot $r4$ and the robotic modules at its right docking interface—at robot $r5$ and $r6$. At robot $r3$, because of the particular endogenous fault, it was not able to donate its battery charge in the organism. But, because of the current limiter at the energy sharing module the on-board electronics was not affected with the particular fault and therefore remained operational. The spikes visible in the Fig. 5.29, occurred because the current limiter at the energy sharing module of each robot periodically tried to re-establish the particular current flow path. In the particular situation, robot $r2$, that was energetically healthier than robot $r1$, donated its battery charge by establishing a sub-power bus with robot $r1$.

Fault Detection and Isolation

Adapting the procedures detailed in algorithm 1 at each robotic module in the particular scenario firstly disabled the energy sharing and successively the current flow

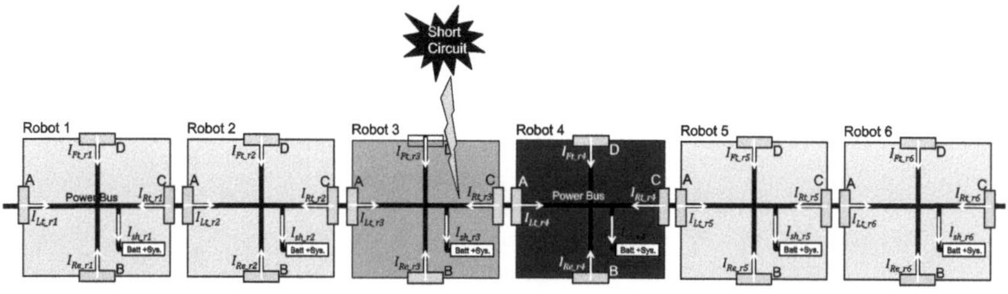

Fig. 5.28 Block diagram highlighting the fault location in the organism

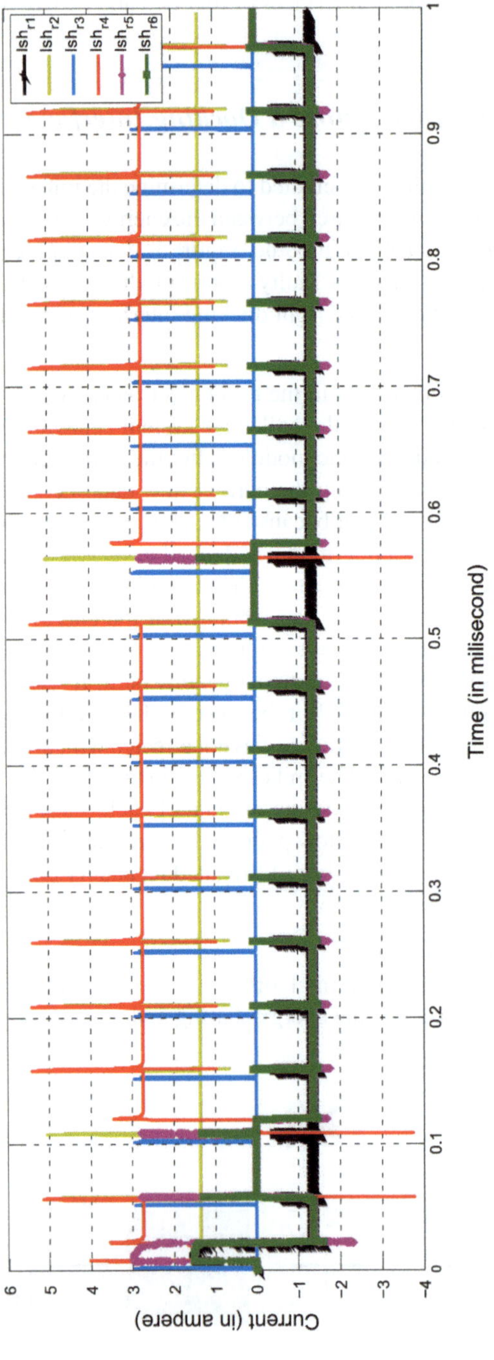

Fig. 5.29 Current flow between the robotic modules in the presence of the injected fault

through the four docking interfaces. To locate and isolate the fault, in the second step, adapting the procedure detailed in algorithm 2 firstly with the disabled docking interfaces enabled the energy sharing at each robotic module in the organism. In the next step, the current flow values through the energy sharing module of each robot was scanned, as if the fault has occurred indigenously. At robot $r3$, on the detection of an endogenous fault condition, it disabled its energy sharing and the four docking interfaces. Whereas, robots $r2$ and $r4$ in response disabled or deactivated their right and left docking interfaces, respectively. Figure 5.30 shows the current flow between the robotic modules after fault detection and isolation. The results show the applicability of fault detection, identification and isolation algorithms that revived power sharing between multiple robotic modules despite a faulty robot in the organism. In effect, the current flow from robot $r4$ to the robotic modules $r5$ and $r6$, and between robot $r2$ and $r1$ is now free from unwanted spikes.

5.4.4 Fault Simulation at an Actuator

The next most probable fault location in a robotic module that can be fatal to a modular robotic organism was considered at an actuator component. For this purpose, a short circuit condition was simulated at an actuator location of a robotic module in the organism. Figure 5.31 graphically highlights the location of a short circuit at an actuator component of robot $r3$.

Fault Injection

To analyze the effect of an actuator failure on the collective behavior of robotic modules, the SPICE simulation was performed after the injection of a short circuit at robot $r3$. Figure 5.32 shows the current flow between the robotic modules in the presence of an actuator fault at robot $r3$. In the particular scenario, the effects of an actuator fault varied greatly from the earlier two scenarios. Notably, the power sharing behavior of robot $r4$ that became paralyzed in the sense that it was not able to donate its on-board energy neither to the robotic modules docked to its right nor to the left docking interface. The particular behavior was because of the omni-directional power bus that allowed an high inrush current to flow from the robot $r4$ to the faulty actuator component of robot $r3$. In effect, the energy sharing module of robot $r4$ disconnected the particular current flow path. On the expiry of a timer, it tried to re-establish the current flow path but failed due to the presence of the particular fault. In addition, the short circuit at the actuator component of robot $r3$ not only ceased its operations but also affected the robotic modules that were not directly docked to it, i.e., the current flow between the robots $r2$ and $r1$, and between $r5$ and $r6$.

Fault Detection and Isolation

Likewise, following the procedure detailed in algorithm 1 and 2 in the current scenario initially turned off energy sharing at each robotic module and successively the isolation of the faulty robot revived power sharing between the unaffected robotic modules of the organism. Figure 5.33 shows the current flow between the robotic

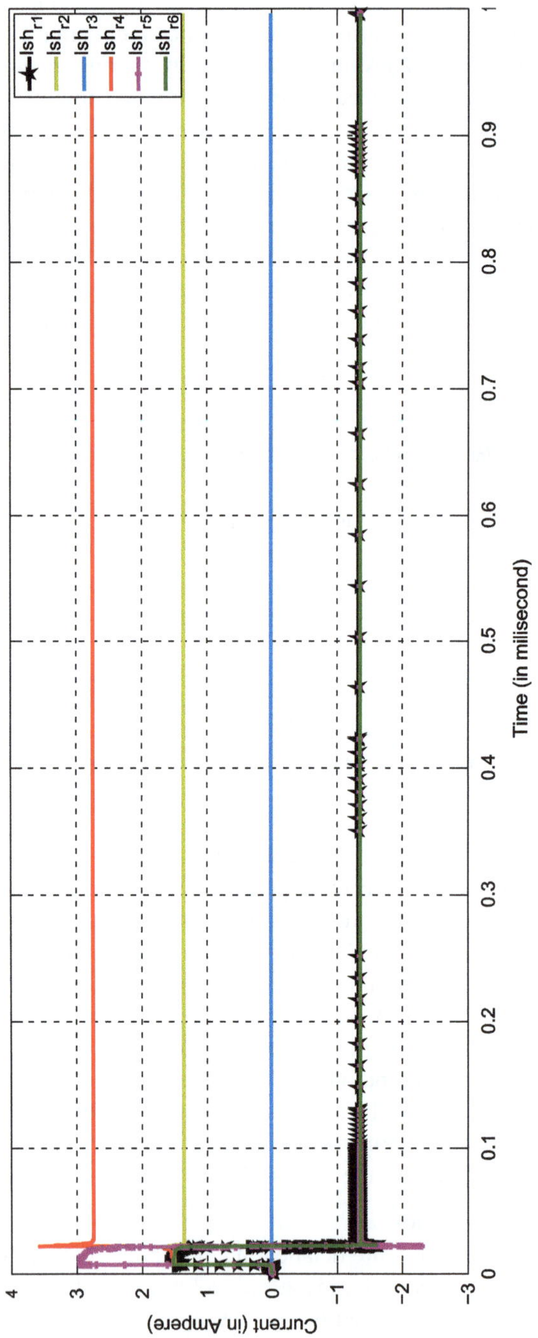

Fig. 5.30 Current flow between the robotic modules after adapting the procedure detailed in algorithm 1 and 2

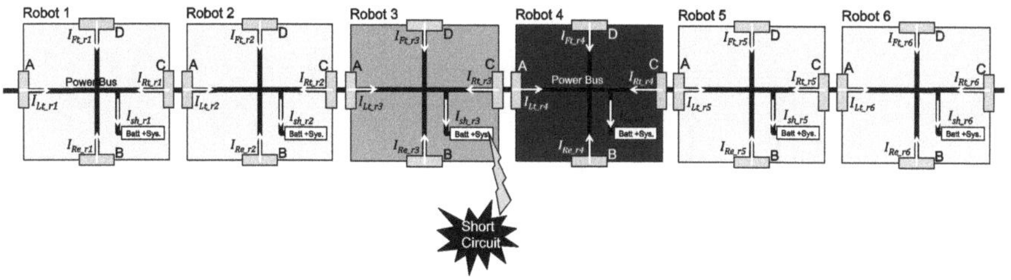

Fig. 5.31 Block diagram highlighting the location of a short circuit at an actuator module of robot $r3$

modules after the detection and isolation of a faulty robot. The fault isolation mechanism at robot $r4$ deactivated its left docking interface, and at robot $r2$, deactivated its right docking interface with robot $r3$. After the isolation of robot $r3$, a sub-power bus was established between $r2$ and $r1$, where $r2$ became the energy donor, and the second sub-power bus between the robots $r4$, $r5$ and $r6$, where $r4$ became the energy donor robot. In other words, the isolation of robot $r3$ in the organism shows that the rest of the robotic modules are now able to share their battery charge with each other and continue their autonomous operations without "splitting"or "reorganization" of the robotic modules.

5.4.5 Sensor Failure Detection

Sensors are always an essential system component in the design of any control system. In regards to the developed power management system, the accuracy of the sensor measurements, employed for measuring the current flow in different system components, determines a system's ability to react in response to endogenous and exogenous perturbations. A sensor failure in a robotic module may arise due to an electrical interference, an improper configuration, a short circuit, power failure, or some external influence that can cause damage to the system electronics.

A sensor fault in the power management system of a robotic module can be detected by exploiting the inherent redundancy between the robotic modules. From the Fig. 5.34, using Kirchhoff's current law, the current flow through the four docking interfaces of robot $r1$ is obtained as

$$I_{Ft\_r1} = I_{Esh\_r1} - (I_{Rt\_r1} + I_{Re\_r1} + I_{Lt\_r1}), \tag{5.9}$$

$$I_{Rt\_r1} = I_{Esh\_r1} - (I_{Ft\_r1} + I_{Re\_r1} + I_{Lt\_r1}), \tag{5.10}$$

$$I_{Re\_r1} = I_{Esh\_r1} - (I_{Ft\_r1} + I_{Rt\_r1} + I_{Lt\_r1}), \tag{5.11}$$

$$I_{Lt\_r1} = I_{Esh\_r1} - (I_{Ft\_r1} + I_{Rt\_r1} + I_{Re\_r1}), \tag{5.12}$$

where, the current flow through the energy sharing module is

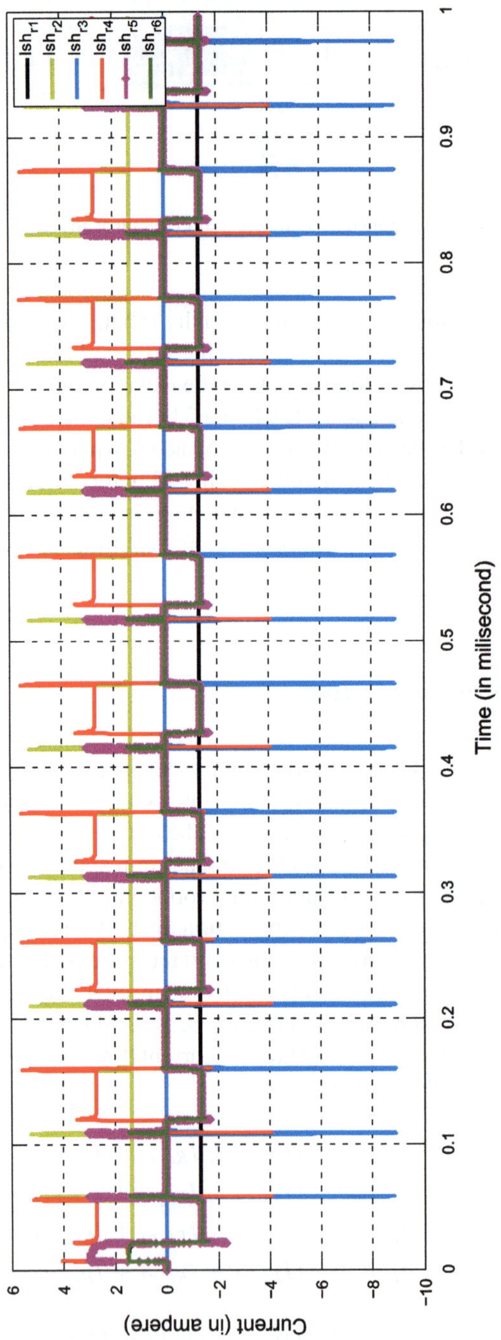

Fig. 5.32 Current flow between the robotic modules in the presence of a short circuit at an actuator component of robot $r3$

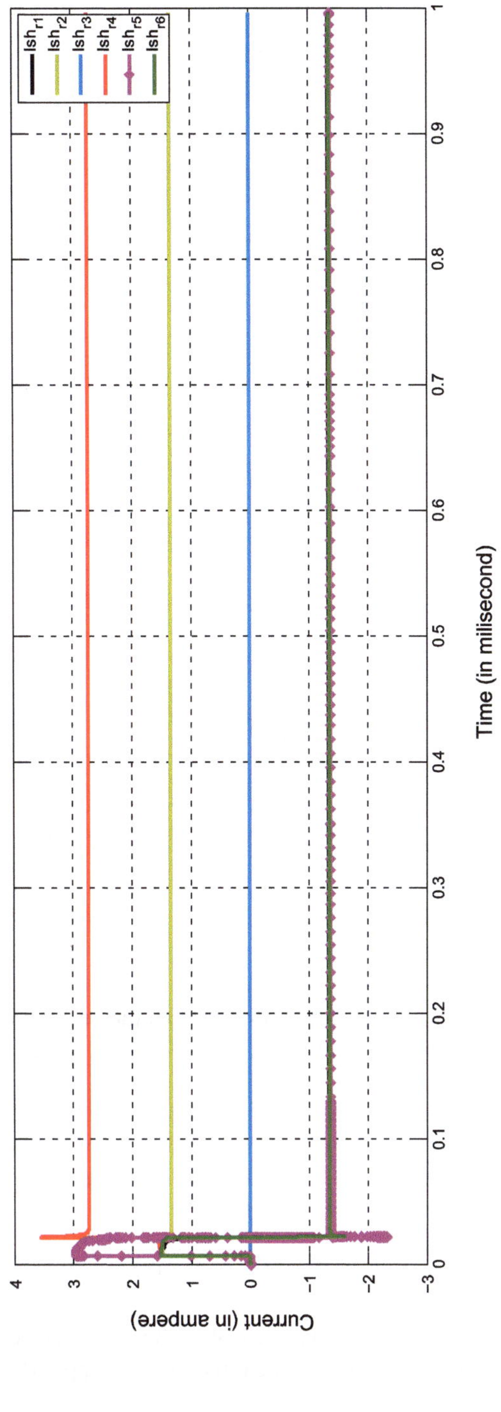

Fig. 5.33 Current flow between the robotic modules after the detection and isolation of a faulty robot

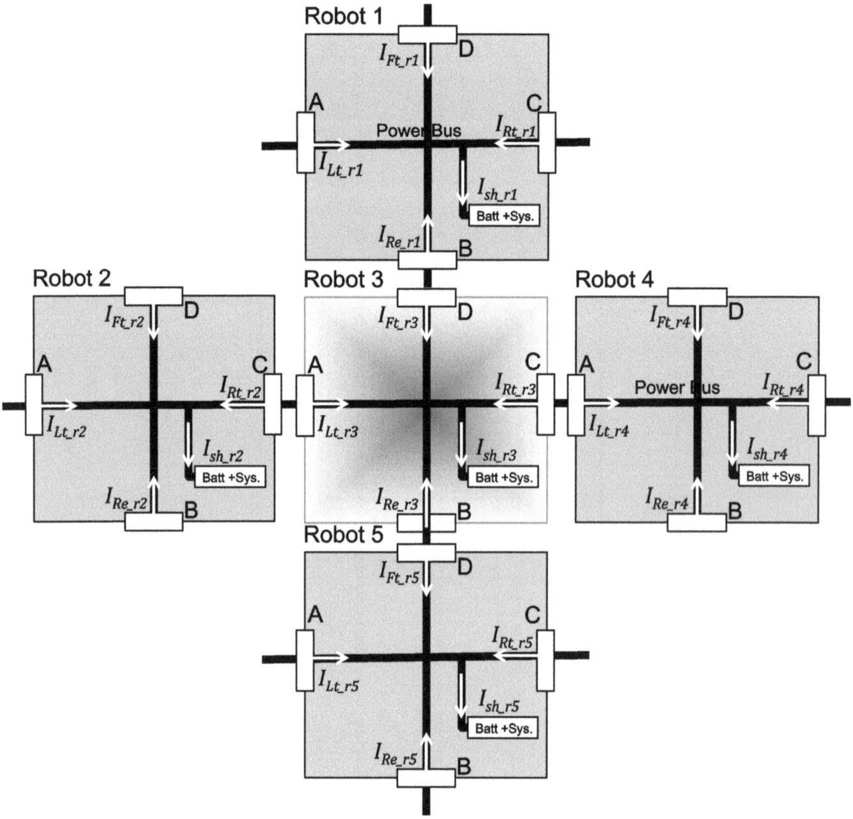

Fig. 5.34 Block diagram of an organism with an omni-directional power bus between the robotic modules. The robot $r3$ is highlighted to show the occurrence of a measurement fault in one of its docking side

$$I_{Esh\_r1} = I_{Batt\_r1} - I_{sys\_r1}. \tag{5.13}$$

The variable I_{Batt_r1} denotes the current drawn from the on-board battery pack and I_{sys_r1} represents the current consumption of the on-board system components of a robotic module.

Figure 5.34 shows the block diagram of an organism with five robotic modules. Let's assume, the homeostatic controller of robot $r3$ encounters a measurement error due to a sensor failure in any one of it's docking side. Using the above equations, the current flow through robot $r3$ is obtained as,

$$I_{Batt\_r3} - I_{sys\_r3} = I_{Ft\_r3} + I_{Rt\_r3} + I_{Re\_r3} + I_{Lt\_r3}. \tag{5.14}$$

A measurement error at any docking interface of a robot can be detected if the equality in Eq. (5.14) does not hold. Using analytical redundancy approach, a sensor failure in one of the docking sides of robot $r3$ can be detected by using the corresponding measurements from its direct neighbors. Now assume the robotic modules $r1$, $r2$, $r4$ and $r5$ in the organism are free from the possible perturbations and only robot $r3$ was suffering with a malfunctioning of a sensor component. In the particular configuration shown in Fig. 5.34, the current flow through the *front* docking side of robot $r3$ must be equivalent to the current flow through the *rear* docking side of robot $r1$, i.e.,

$$\begin{aligned} I_{Ft\_r3} &= I_{Re\_r1}, \\ I_{ESh\_r3} - (I_{Rt\_r3} + I_{Re\_r3} + I_{Lt\_r3}) &= I_{ESh\_r1} - (I_{Ft\_r1} + I_{Rt\_r1} + I_{Lt\_r1}). \end{aligned} \tag{5.15}$$

Similarly, the current flow through the *right* docking sides can be compared with,

$$\begin{aligned} I_{Rt\_r3} &= I_{Lt\_r4}, \\ I_{ESh\_r3} - (I_{Ft\_r3} + I_{Re\_r3} + I_{Lt\_r3}) &= I_{ESh\_r4} - (I_{Ft\_r4} + I_{Rt\_r4} + I_{Re\_r4}), \end{aligned} \tag{5.16}$$

through the *rear* docking side of robot $r3$ is compared with,

$$\begin{aligned} I_{Re\_r3} &= I_{Ft\_r5}, \\ I_{ESh\_r3} - (I_{Ft\_r3} + I_{Rt\_r3} + I_{Lt\_r3}) &= I_{ESh\_r5} - (I_{Rt\_r5} + I_{Re\_r5} + I_{Lt\_r5}), \end{aligned} \tag{5.17}$$

and, finally through the *left* docking side of robot $r3$ is compared with the right docking side of robot $r2$, i.e.,

$$\begin{aligned} I_{Lt\_r3} &= I_{Rt\_r2}, \\ I_{ESh\_r3} - (I_{Ft\_r3} + I_{Rt\_r3} + I_{Re\_r3}) &= I_{ESh\_r2} - (I_{Ft\_r2} + I_{Re\_r2} + I_{Lt\_r2}). \end{aligned} \tag{5.18}$$

For the detection of an occurred measurement fault, if the following equality holds,

$$I_{ESh\_r3} = I_{Re\_r1} + I_{Lt\_r4} + I_{Ft\_r5} + I_{Rt\_r2}, \tag{5.19}$$

then a sensor failure in the corresponding docking side can be easily detected.

5.4.6 Limitation of the Simulation System

In literature several simulation platforms have been developed and used to conduct behavior based simulations but non of them in fact integrated the control mechanisms, e.g., at the application layer, with circuit simulation programs, like, SPICE.

Since it was not defined in the scope of this work, the main focus was only laid to develop and analyze the distributive fault tolerance mechanisms at the application layer of the physically docked robotic modules. The simulation program used for the experiments only provides circuit based simulation. To further explore the distributive fault detection, identification and isolation algorithms and their effect on the behavior of a modular robotic organism, a simulation framework is required that not only performs SPICE simulations but also allows to program the algorithms.

5.5 Summary

In the chapter, in the beginning, the experiment section explored the operational behavior of the proposed power management system with fault tolerant energy sharing, implemented and integrated in the design of REPLICATOR robotic modules, during artificial trophallaxis, power sharing, and in an emulated fault situation. In the later half of the chapter, the devised simulation framework—*Replicator Power Flow Simulator*—was then used to simulate and explore the effects of different forms of power sharing on the operations of multiple robotic modules in two exemplary organism morphologies. From the perspective of fault tolerance, to investigate the collective behavior before and after the fault isolation at an organism level, the experimental section firstly presented the simulation setup. In the simulation experiments an exogenous and two endogenous fault conditions in the robot electronics were introduced to observe their affect on the power sharing between multiple robotic modules in the organism. In addition, by exploiting the inherent redundancy in the power management system a simple mechanism was shown to detect a sensor failure at a robotic module in an organism.

Chapter 6
Conclusion and Outlook

6.1 Conclusion

In this research work, the principles of self-sufficiency defined earlier for a standalone robot, have been explored and developed from the perspective of an autonomous re-configurable modular robotic organism. It this regard, firstly, the open issues at a modular robotic organism level that arise in different scenarios are described. To address the described issues or challenges, a behavior modeling approach was used to define the self-sufficient behavior of an autonomous modular robotic organism. The proposed behavioral model further highlighted the challenges involved in different behavioral states and conditions under which the multiple robotic modules in an organism are required to synchronize and adapt their individual actions to achieve energetic autonomy. In the other words, the condition under which an organism has to adapt different behavioral states to remain self-sufficient. More specifically, from the perspective of self-sufficiency, the behavioral states in which an organism will be required to adapt different modes of power sharing between the robotic modules. For instance, a *nesting* organism may freely use the available energetic resources to utilize all on-board tools to explore the environment or to accomplish different tasks in the arena. On the other hand, the same organism in *searching* or *partial breakdown* states may not adapt the similar power sharing topology among the robotic modules due to different reasons. In this regard, to address the issues involved in power sharing between the robotic modules of an organism, a three layered architecture of a modular robotic organism is used to present a novel dynamic power management system.

The proposed power management system functionalities were then divided among the three layers, i.e., hardware, middleware and the application software layer. At first, a novel power management system design with fault tolerant energy sharing at the hardware layer was proposed and developed. The proposed power management system was then integrated in the design of a re-configurable robotic platform to test its performance under different conditions. The experiment results verified the innate features, power sharing and fault tolerance, of the proposed system during artificial

R.H. Qadir, *Self-Sufficiency of an Autonomous Reconfigurable Modular Robotic Organism*, Adaptation, Learning, and Optimization 17,
DOI 10.1007/978-3-319-10289-4_6

trophallaxis and power sharing in a modular robotic organism. The applicability of artificial trophallaxis in a battery powered robot swarm greatly depends on the specific robot platform design, which includes battery chemistry, docking contacts for power transfer and the charging mechanism. The trophallaxis efficiency that varied between 82–85 % with the proposed power management system is therefore used in the design of real robotic modules to further evaluate its effectiveness in a robot swarm under different operating conditions. The characteristic features of the proposed power management system, beneficial for the dynamic operations of a modular robotic organism, verified by performing the experiments are the dynamically controllable power sharing and innate fault tolerance. The implementation of these two features at the hardware layer in fact proves that the set goals are achieved at a single robotic module level.

For the implementation of the proposed application software layer power management components a simulation framework called *Replicator Power Flow Simulator* was then devised due to the unavailability of a sufficient number of real robotic modules. The simulation framework provided a platform to examine the power sharing behavior of modular robotic organisms that no other simulation tool offered at the time of writing this work. Due to the omnipresent risk of individual robot breakdown, static power sharing is no option because it cannot cope with faulty robots in the organism. Therefore, despite the higher complexity compared to static power sharing, a dynamic power sharing strategy needs to be implemented in order to achieve the goal of an robust and long-living organism. As already stated, the focal point behind the simulation experiments was to obtain alternate power sharing mechanisms that can provide an equivalent amount of operational time in different scenarios. The analysis of the simulation results, where the effectiveness and applicability of different power sharing policies was observed, also revealed the different inter-dependent factors involved in the self-sufficiency of a modular robotic organism, as mentioned before. With *static power sharing*, the robotic modules in the organism shared their charge with each other even when some of them did not require external power. Because of this, the energetically healthy robotic modules had to compromise their self-sufficiency in the organism. Alternatively, the *controlled power sharing* strategies allowed the robotic modules to share their charge with each other, but only during the time it was required. In comparison, the devised dynamic power sharing policies provide more flexibility in terms of energy sharing and therefore can be chosen on the fly by an autonomous modular robotic organism with the changing behavioral states. For instance, in a partial breakdown behavioral state when static power sharing topology is not possible, an organism using dynamic power sharing topology can accomplish its objective without splitting or reorganizing the robotic modules, that in addition requires energetic resources. In terms of advantages, as stated earlier, the robotic modules in the organism with *dynamic power sharing* policies shared their on-board battery charge only during the time it was required. In this manner, the energetic autonomy of the individual robotic modules remained preserved locally that in case of a fault or breakdown situation may allow the robotic modules to adapt a new morphology or operate as standalone entities, after detaching themselves from the organism. Another significant advantage of the devised power sharing strategies is that it empowers the robotic modules to form sub-power buses to share power with each other despite faulty robots in the organism, explored in fault tolerance simulation experiments.

In the next step, the application software layer fault tolerance feature of the proposed power management system was explored using a SPICE simulation tool. The simulation results evidently showed that the impact of a fault or a component failure in a modular robotic organism in fact depends on the three factors: *fault type*, *position* of the faulty robot in an organism's morphology and its *role* or *energetic status* in the organism. Considering the simulation results, in the first scenario, the aftereffects of an exogenous fault, occurring between the docking interfaces of two robotic modules, were observed as less severe than the two endogenous fault conditions. That is, the particular exogenous fault only affected the robotic modules that were directly connected to the fault location. The particular behavior was because of the power management system design which stopped the occurred fault to introduce deviations in the system operations. In the other two endogenous fault conditions, in the absence of any application layer fault tolerance mechanism, the system's innate fault tolerance resisted the fault condition but cannot revive the steady state system operations. The simulation results showed that adapting the procedure defined in the two proposed algorithms allowed the robotic modules to revive their steady state system operations—power sharing—in the presence of endogenous and exogenous faults.

At the end, the combination of the two components of artificial energy homeostasis, i.e., controlled power sharing and distributed fault tolerance, proposed to achieve self-sufficiency is expected to increase the robustness and stability of a real modular robotic organism, to operate and survive in different circumstances.

6.2 Outlook

Often a research project opens new dimensions and raises several more questions than it answers. This research work on self-sufficiency of an autonomous self-reconfigurable modular robotic organism, where different solutions to different inter-linked problems have been explored and developed, still demand further exploration and development on some topics.

6.2.1 Implementation of the Proposed Algorithms on Real Robotic Modules

Apart from the simulation environment, it is equally important to implement the proposed algorithms on real robotic modules to gain more experience and knowledge about their applicability and effectiveness in different operating conditions.

In this regard, using multiple parameters, e.g., energy sharing threshold, initial energy distribution among the robotic modules, locomotion pattern, etc., the implementation of artificial energy homeostasis can be improved. That is, adaptation of newly learned set points, physiological limits, and regulatory responses for a more efficient power sharing. Also, the application of fault detection and identification, and fault isolation algorithms especially during power sharing can be explored to further improve their effectiveness in different organism morphologies. For this purpose, the arena conditions and the operating environment must be specified to evaluate the effectiveness of combination of different strategies.

6.2.2 Behavioral Modeling

From the perspective of behavioral modeling, a possible extension to the existing macroscopic behavioral model of a modular robotic organism could be the inclusion of different interdependent parameters in a mathematical model. These parameters may include kinematics, internal state, position of a robotic module in the organism's structure, and its role or responsibility in the organism. A well defined mathematical model will certainly help to model and analyze the emergent behavior of multiple robotic modules docked in different configurations in an organism. Another possible extension is to evaluate the quality of models by comparing them against the real hardware.

6.2.3 Improved Simulation System

To further explore the challenges of self-sufficiency at a modular robotic organism level, another possible extension in the current work could be the integration of the *Replicator power flow simulator* (see Sect. 4.3) in a simulation framework that can simulate the power flow and the physical movements of a modular robotic organism in 3D space. One possible option can be to use the open source Robot3D simulator (Winkler and Wörn 2009; Winkler et al. 2012). It is proposed because the Robot3D simulator used the design and form factor of the REPLICATOR robotic modules to dynamically simulate a swarm of mobile robots and modular robotic organisms. More interestingly, it provides the movements of a swarm of robots, simulation of the docking procedure between the robotic modules and the organism locomotion in a 3D environment. Such an enriched simulation framework in result can also be used to further explore and develop the fault tolerance at a modular robotic organism level.

Appendix A
Fault Identification at an Individual Level

Faults are commonplace in mobile robotic platforms and the novel electrical and mechanical design of the REPLICATOR platform amplifies this problem. A range of faults can be expected, from those caused by the behavioral controllers to those effecting particular sensors or actuators and even those originating from the power management system itself.

For the implementation of a fault tolerant system, firstly, it is important to differentiate the different kinds of faults that can develop in critical system components and bear tendency to bring the whole system to collapse. Using the fault tree analysis (FTA) approach, a fault tree identifying different sources with probable failures at an individual robotic module level was developed using a REPLICATOR robotic module as an example (see Fig. A.1).

The top event represents the failure of an entire robotic system. The primary causes of a robot failure can originate from different sensors, power system, communication system, controllers, actuators or some external influence.

- **Sensor failure**

  Sensors are always an integral part of any control system both in biological and artificial systems. A standalone autonomous robotic module usually requires a variety of sensors to keep its autonomy under different operating conditions. The different types of sensors may include cameras, lasers, infra-red light detectors, gyroscopes, current sensors, sensors for controlling actuators, etc. These sensors on one side provide stability and increase reliability of a control system in case of a malfunction can directly and indirectly cease its operations. Figure A.2 shows the fault tree aggregating different causes of a sensor failure at an individual robot level. The different types of faults include noisy output, an improper setup, a short circuit, power failure, or some external influence in the form of a hardware damage. A noisy measurement from a sensor may result either due to an improper setup, interference or an internal failure.

R.H. Qadir, *Self-Sufficiency of an Autonomous Reconfigurable Modular Robotic Organism*, Adaptation, Learning, and Optimization 17,
DOI 10.1007/978-3-319-10289-4

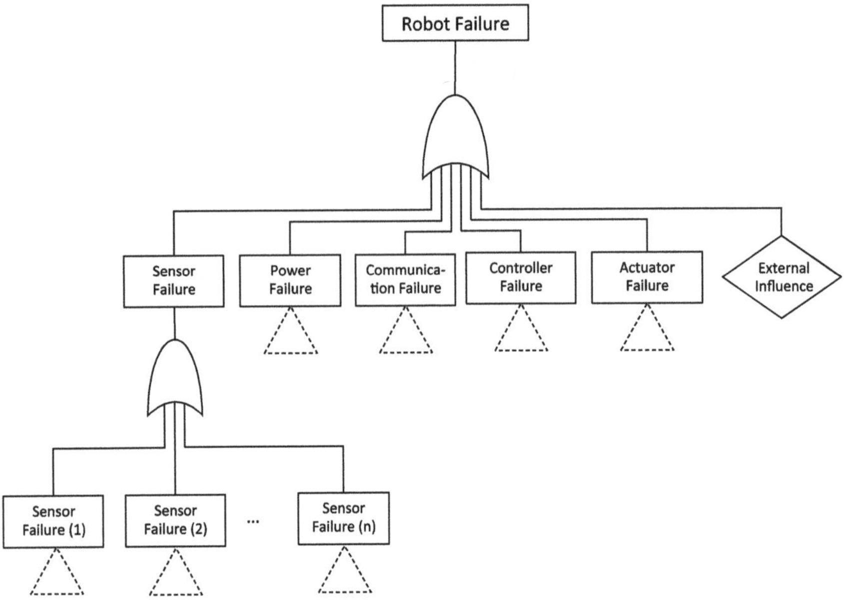

Fig. A.1 Fault tree combining failures from different system components that may lead to a robot failure

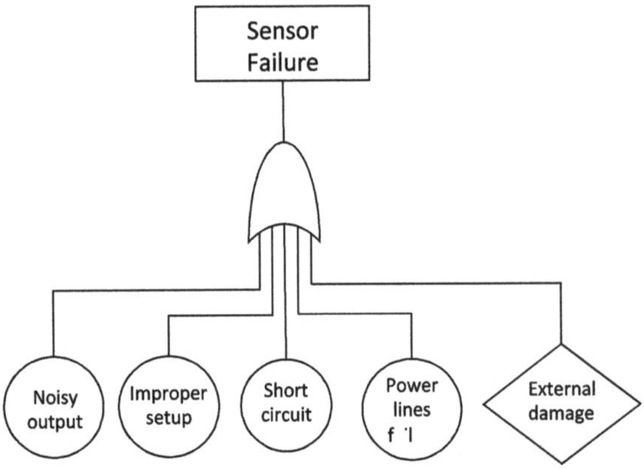

Fig. A.2 Sensor failure fault tree at an individual robot level

- **Power failure**

  An uninterrupted power supply is vital to keep the operations of an artificial system alive. Considering the electro-mechanical design of a REPLICATOR robotic module a fault in the system that may lead to a complete power failure can originate from different system components. Figure A.3 shows the fault tree combining different types of failures that can contribute to a total power failure. The power breakdown in a robotic module can occur mainly because of a short circuit in the

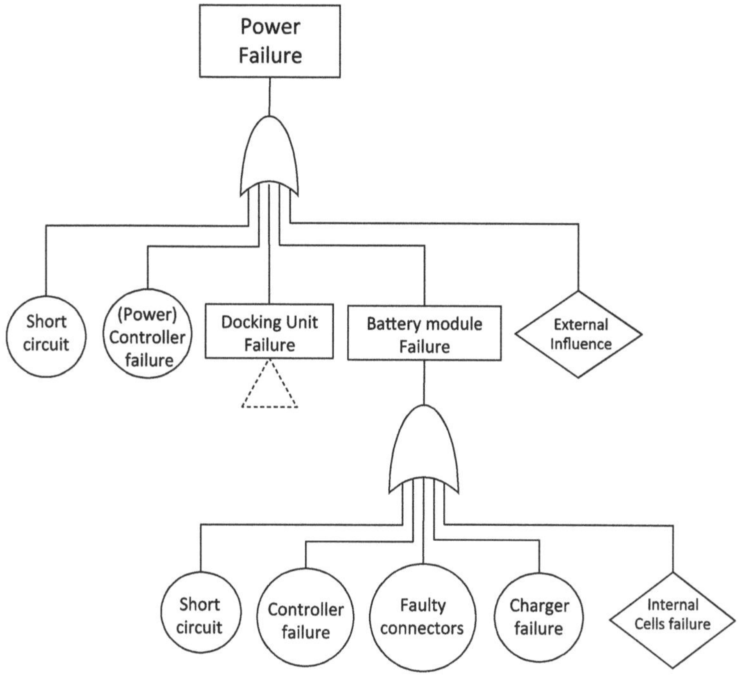

Fig. A.3 Power failure fault tree at an individual robot level

robot electronics, a failure in the power management controller, at docking interfaces, in the battery management module or some external influence. The factors involved in the failure of a *battery management module* include a short circuit, battery controller failure, faulty connectors or terminals, charger failure or some internal factors, e.g., dead cells, deep discharged cells, etc.

- **Communication failure**

  A robust communication mechanism serves as a backbone for establishing and synchronizing the group activities in a robot swarm. The communication mechanism usually comprises buses and transceivers to connect the control system with the sensors and effectors. Therefore, a failure in a component responsible for intra-robot communication may indirectly lead to a robot failure. Figure A.4 shows the fault tree combining the different sources that can contribute to a communication failure. It includes noise or interference, an improper start up configuration of a communication protocol, a short circuit, a broken wired link, or some component damage.

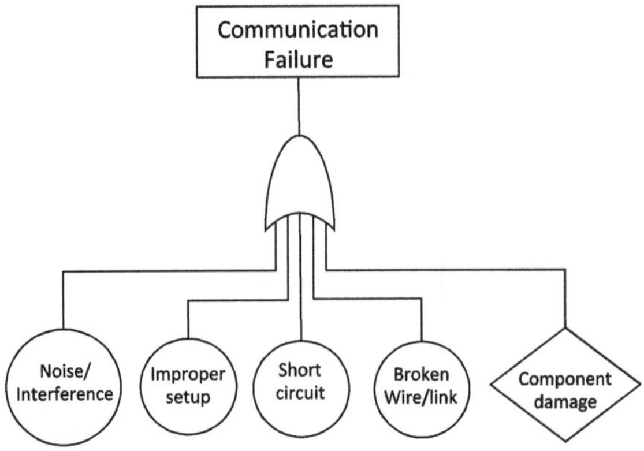

Fig. A.4 Communication failure fault tree at an individual robot level

- **Actuator failure**

  Actuators are always an essential system component to maintain the autonomy of an autonomous mobile robotic system. Actuator failures generally refer to those malfunctions that are linked with the motor drive(s) of a mobile robot. Figure A.5

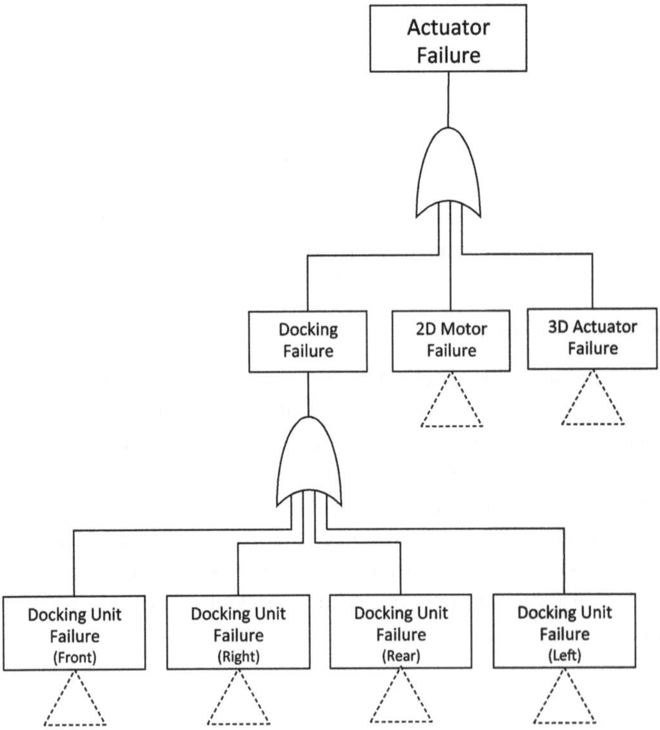

Fig. A.5 Actuator failure fault tree at an individual robot level

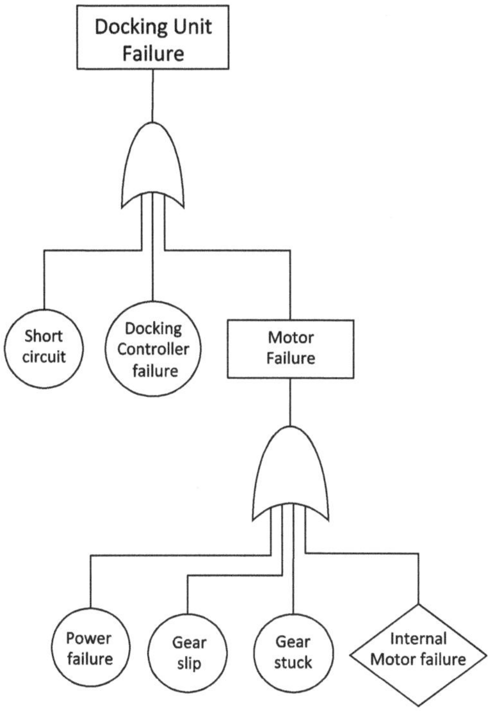

Fig. A.6 Docking failure fault tree at an individual robot level

shows the fault tree highlighting different components that can contribute to an actuator failure. From the design of a REPLICATOR robotic module, the causes of an actuator failure can originate either from one of the four docking interfaces, 2D locomotion drives or a 3D actuator.

The docking interface failures are linked with the docking units, on each lateral side of a robotic module for intra-organism power sharing and wired communication. Figure A.6 shows the fault tree highlighting the possible failure causes that can contribute to a faulty docking unit. The probable causes include a short circuit at the docking interface pins of a docking unit, docking controller failure, or a docking motor drive failure. A *docking motor failure* may results because of a power failure, a gear slip, a gear stuck or some internal motor failure.

The 2D motor unit is relatively a simple module but may produce a failure because of a short circuit, a power failure or some internal motor drive failure. Figure A.7a shows the fault tree of a 2D motor drive. And, Fig. A.7b shows the fault tree of a 3D actuator drive. A failure in the 3D actuator drive may be caused by controller failures, stuck gear, a short circuit, a power failure or some internal malfunctioning.

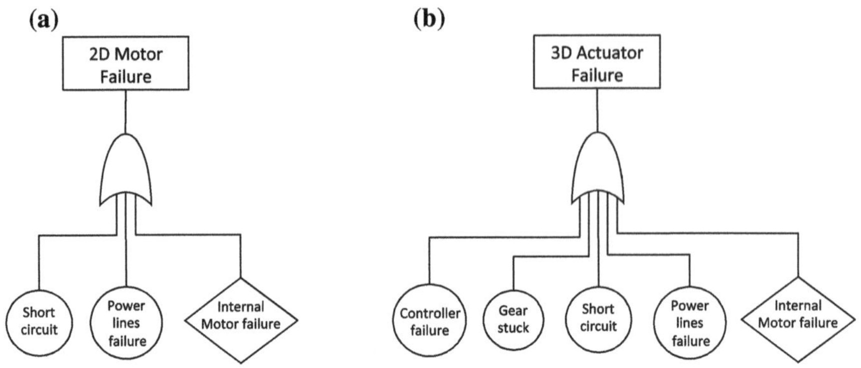

Fig. A.7 Actuator fault tree: **a** 2D Motor failure fault tree, **b** 3D actuator failure fault tree at an individual robot level

- **Controller failure**

A robot's control unit, implemented as one or more microcontrollers or microprocessors, is usually less prone to hardware failures compared to other electronic components. Figure A.8 shows the fault tree combining the possible causes that may result in a controller failures. A control unit usually experiences failure because of a defective port, a software bug or a hardware failure, e.g., a faulty crystal, improper soldering, physical damage, etc.

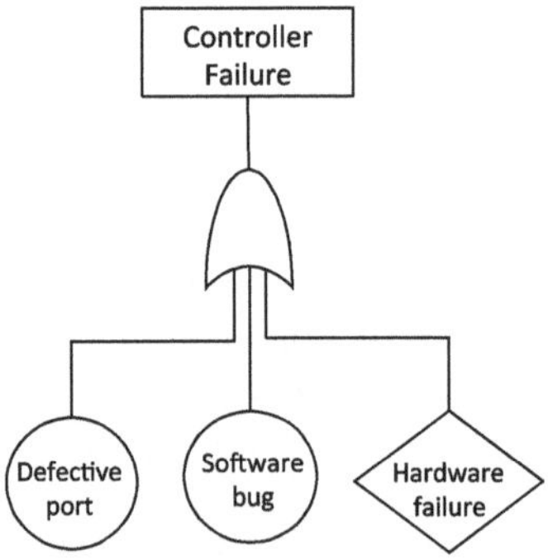

Fig. A.8 Controller failure fault tree at an individual robot level

References

W.R. Ashby, *Design for a Brain: The Origin of Adaptive Behaviour*, 2nd edn. (Chapman and Hall, London, 1960)

K. Astrom, B. Wittenmark, *Adaptive Control* (Addison Wesley, Boston, 1989)

E. Bahgeçi, E. Sahin, Evolving aggregation behaviors for swarm robotic systems: A systematic case study, in *Proceedings of IEEE Swarm Intelligence Symposium*, 2005. SIS 2005, pp. 333–340. IEEE, 2005

L. Bayindir, E. Şahin, A review of studies in swarm robotics. Turk. J. Electr. Eng. Comput. Sci. **15**, 115–147 (2007)

R. Beckers, O.E. Holland and J.L. Deneubourg. From local actions to global tasks: Stigmergy and collective robotics, in *Proceedings of the 4th International Workshop on the Synthesis and Simulation of Living Systems, Artificial Life IV*, ed. by Rodney A. Brooks, Pattie Maes (MIT Press, Cambridge, July 1994), pp. 181–189

G. Beni and J. Wang. Swarm intelligence in cellular robotic systems, in *NATO Advanced Workshop on Robotics and Biological Systems*, June 1989

L. Benini, G. Micheli, *Dynamic Power Management: Design Techniques and CAD Tools* (Kluwer Academic Publishers, Norwell, 1998)

C. Bernard, A. Dastre, *Leçons sur les phénomènes de la vie, communs aux animaux et aux végétaux*, vol. 2 (J.-B, Baillière, 1879)

J.M. Bishop. Stochastic searching networks, in *1st IEE Conference on Artificial Neural Networks*, pp. 329–331, London, 1989

A. Boletis, W. Driesen, J.-M. Breguet, A. Brunete, Solar cell powering with integrated global positioning system for mm3 size robots, in *IEEE/RSJ International Conference on Intelligent Robots and Systems, (IROS'06)*, pp. 5528–5533, October 2006. doi:10.1109/IROS.2006.282191

E. Bonabeau, C. Meyer, Swarm Intelligence: A Whole New Way To Think About Business. Harvard Business Review, Harvard Business School Publication Corporation, Boston, (2001)

E. Bonabeau, M. Dorigo, G. Theraulaz, *Swarm Intelligence: From Natural to Artificial Systems* (Oxford University Press, New York, 1999)

B.J. Cabrera, M.K. Rust, Caste differences in feeding and trophallaxis in the western drywood termite, incisitermes minor (hagen) (isoptera, kalotermitidae). Insectes Soc. **460**(3), 244–249 (1999)

S. Camazine, The regulation of pollen foraging by honey bees: how foragers assess the colony's need for pollen. Behav. Ecol. Sociobiol. **32**, 265–272 (1993)

W.B. Cannon, Physiological regulation of normal states: some tentative postulates concerning biological homeostatics. Jubilee Volume for Charles Richet, pp. 91–93, (1926)

W.B. Cannon, Organization for physiological homeostasis. Physiol. Rev. **9**, 399–431 (1929)

W.B. Cannon, *The Wisdom of the Body* (W.W. Norton, New York, 1932)

R.H. Qadir, *Self-Sufficiency of an Autonomous Reconfigurable Modular Robotic Organism*, Adaptation, Learning, and Optimization 17,
DOI 10.1007/978-3-319-10289-4

Y.U. Cao, A.S. Fukunaga, A.B. Kahng, Cooperative mobile robotics: antecedents and directions. Autonomous Robots **4**, 226–234 (1997)

J. Carlson, R.R. Murphy, How ugvs physically fail in the field. Robot. IEEE Trans. 210 (3):0 423–437, June 2005. ISSN 1552–3098. doi:10.1109/TRO.2004.838027

E.L. Charnov, Optimal foraging, the marginal value theorem. Theor. Popul. Biol. **9**(2), 129–136 (1976)

Y. Chih-Han, N. Radhika, Self-adapting modular robotics: A generalized distributed consensus framework. 2009. http://citeseerx.ist.psu.edu/viewdoc/summary?doi=10.1.1.163.1815

V.G. Claudia, Z. Shlomo, Decentralized control of cooperative systems: categorization and complexity analysis. J. Artif. Intell. Res. **22**, 143–174 (2004)

I.D. Couzin, J. Krause, R. James, G.D. Ruxton, N.R. Franks, Collective memory and spatial sorting in animal groups. J. Theor. Biol. **218**, 1–11 (2002)

E. Şahin, Swarm robotics: from sources of inspiration to domains of application, in *Swarm Robotics Workshop: State-of-the-Art Survey*, Lecture Notes in Computer Science (LNCS), ed. by E. Şabin, W. Spears (Springer, New York, 2005), pp. 10–20

F. Cynthia, Failure recognition and fault tolerance of an autonomous robot. Adapt. Behav., 20 4(0), pp. 375–398, March 1994. ISSN 1059–7123. doi:10.1177/105971239400200403

K.W. Dailey, *The Failure Mode and Effects Analysis Handbook* (DW Publishing company, Dearborn, 2004)

M. Dorigo, Swarm-bots and swarmanoid: Two experiments in embodied swarm intelligence, in *Proceedings of the 2009 IEEE/WIC/ACM International Joint Conference on Web Intelligence and Intelligent Agent Technology*, volume 02 of WI-IAT '09, pp. 2–3, Washington, DC, USA, 2009. IEEE Computer Society

M. Dorigo, G.D. Caro. Ant colony optimization: a new meta-heuristic, in P.J. Angeline, Z. Michalewicz, M. Schoenauer, X. Yao and Ali Zalzala, ed., *Congress on Evolutionary Computation (CEC'99)*, pp. 1470–1477 (Piscataway, NJ, IEEE Press, 1999)

M. Dorigo, E. Şahin, Swarm-robotics - special issue. Autonomous Robots **17**, 111–113 (2004)

M. Dorigo, V. Maniezzo, A. Colorni, Positive feedback as a search strategy. Technical Report 91–016, Dipartimento di Elettronica, Politecnico di Milano, Milan, 1991

M. Dorigo, V. Maniezzo, A. Colorni, Ant system: optimization by a colony of cooperating agents. IEEE Trans. Syst. Man Cybern. Part B **26**, 29–41 (1996)

G.J. Driessen, Cause tree analysis, measuring how accidents happen and the probabilities of their cause. Technical report,Presented to American Psychological Association, Miami, FL, USA, Sep. 1970

G. Dudek, E. Jenkin, D. Wilkes, A taxonomy for swarm robots, in *Proceedings of IEEE/RSJ International Conference on Intelligent Robots and Systems (IROS'93)*, vol. 1 (Yokohama, Japan, 1993), pp. 441–447

T. Eisner, W.L. Brown, The evolution and social significance of the ant proventriculus, in *Proceedings Tenth International Congress of Entomology***2**, 503–508 (1956)

T. Eisner, E.O. Wilson, Radioactive tracer studies of food transmission in ants, in *Proceedings Tenth International Congress Entomology***2**, 509–514 (1958)

A.E. Emerson, Regenerate behavior and social homeostasis of termites. Ecology **37**, 248–258 (1956)

C. Ferrell, *Robust Agent Control of an Autonomous Robot with Many Sensors and Actuators* (MIT, Cambridge, 1993)

M. Franois, *Soft Computing in Engineering Design and Manufacturing*, Chapter Adaptability by Behavior Selection and Observation for Mobile Robots (Springer, New York, 1997)

T. Fukuda, S. Nakagawa, Approach to the dynamically reconfigurable robotic system. J. Intell. Rob. Syst. **1**, 55–72 (1988)

F.J. Garcıa, V. Izquierdo, L.J. de Miguel, J.R. Perán, Fault-diagnostic system using analytical fuzzy redundancy. Eng. Appl. Artif. Intell. **13**(4), 441–450 (2000)

S. Garnier, J. Gautrais, G. Theraulaz, The biological principles of swarm intelligence. Swarm Intell. **1**, 3–31 (2007)

J. Gautrais, P. Michelena, A. Sibbald, R. Bon, J.-L. Deneubourg, Allelomimetic synchronisation in merino sheep. Anim. Behav. **74**(5), 1443–1454 (2007)

B. Gerkey, R. Vaughan, A. Howard, The player/stage project: Tools for multi-robot and distributed sensor systems, in *11th International Conference on Advanced Robotics (ICAR 2003)*, pp. 317–323, Coimbra, Portugal, June 2003. http://citeseer.ist.psu.edu/gerkey03playerstage.html

J. Gertler, *Fault Detection and Diagnosis in Engineering Systems* (CRC, New York, 1998)

D. Goldberg, M. J. Matarić, Robust behavior-based control for distributed multi-robot collection tasks. CiteSeerX - Scientific Literature Digital Library (United States), 2000. http://citeseerx.ist.psu.edu/viewdoc/summary?doi=?doi=10.1.1.37.3594

P.J.S. Goncalves, P. J.D. Torres, C.M.O. Alves, F. Mondada, M. Bonani, X. Raemy, J. Pugh, C. Cianci, A. Klaptocz, S. Magnenat, J.C. Zufferey, D. Floreano, A. Martinoli, *in Proceedings The e-puck, a robot designed for education in engineering* (2009)

D.M. Gordon, The organization of work in social insect colonies. Nature **380**, 121–124 (1996)

D.M. Gordon, Control without hierarchy. Nature **446**(7132), 143 (2007)

S. Goss, S. Aron, J.L. Deneubourg, J.M. Pasteels, Self-organized shortcuts in the Argentine ant, in *Naturwissenschaften*, vol. 76, pp. 579–581. Springer-Verlag, 1989

A.M. Green, K. Emery, Y. Hishikawa, W. Warta, Solar cell efficiency tables (version 33). Prog. Photovoltaics Res. Appl. **17**, 85–94 (2009)

W. Grey, *The Living Brain* (W. W. Norton, New York, 1963)

C. Grosan, A. Abraham, M. Chis. Swarm intelligence in data mining, in *Studies in Computational Intelligence (SCI)*, vol. 34, pp. 1–20, Heidelberg, 2006. Springer-Verlag, Berlin

D. Grúnbaum, S.V. Viscido, J.K. Parrish eds., Extracting interactive control algorithms from group dynamics of schooling fish, in Heidelberg Springer, in *Proceedings of the cooperative control conference*, pp. 103–117, 2005

C. Hamilton, B.T. Lejeune, R.B. Rosengaus, Trophallaxis and prophylaxis: social immunity in the carpenter ant camponotus pennsylvanicus. Biol. Lett. **7**(1), 89–92 (2010)

A.T. Hayes, P. Dormiani-Tabatabaei, Self-organized flocking with agent failure: Off-line optimization and demonstration with real robots, in *IEEE International Conference on Robotics and Automation, 2002. Proceedings*. ICRA'02., vol. 4, pp. 3900–3905. IEEE, 2002

D. Helbing, P. Molnár, I.J. Farkas, K. Bolay, Self-organizing pedestrian movement. Environ. Plann. B: Plann. Des. **28**(3), 361–383 (2001)

D.M. Himmelblau, *Fault Detection and Diagnosis in Chemical and Petrochemical Processes* (Elsevier, Oxford, 1978)

A. Houston, P. Schmid-Hempel, A. Kacelnik, Foraging strategy, worker mortality and the growth of the colony in social insects. Am. Nat. **131**, 107–114 (1988)

A. Howard, M.J. Matarić, G.S. Sukhatme, An incremental self-deployment algorithm for mobile sensor networks. Auton. Robots **13**(2), 113–126 (2002a)

A. Howard, M.J. Mataric, G.S. Sukhatme, Mobile sensor network deployment using potential fields: A distributed, scalable solution to the area coverage proble, in *Proceedings of the 6th International Symposium on Distributed Autonomous Robotics Systems (DARS02)*, pp. 299–308. Citeseer, 2002

R. Humza, O. Scholz, Energy autonomy and energy harvesting in reconfigurable swarm robotics, in P. Levi, S. Kernbach eds., *Symbiotic Multi-Robot Organisms*, pp. 116–135. Springer-Verlag, 2010. ISBN 978-3-642-11691-9

R. Humza, O. Scholz, A case study on self-sufficiency of individual robotic modules in an arena with limited energy resources, in *ADAPTIVE 2011, The Third International Conference on Adaptive and Self-Adaptive Systems and Applications*, pp. 29–35, 2011

C.I-Ming, Y. Guilin, K. In-Gyu, Numerical inverse kinematics for modular reconfigurable robots. Journal of Robotic Systems, CCC 0741–2223/ 99/ 040213–13: 213–225, Nov. 1999

I. Ieropoulos, C. Melhuish, J. Greenman, Artificial metabolism: Towards true energetic autonomy in artificial life, eds. by W. Banzhaf, J. Ziegler, T. Christaller, P. Dittrich, J. Kim, *Advances in Artificial Life*, Lecture Notes in Computer Science, vol. 2801 (Springer, Berlin / Heidelberg, 2003) pp.792–799 ISBN 978-3-540-20057-4

I.A. Ieropoulos, J. Greenman, C. Melhuish, J. Hart, Comparative study of three types of microbial fuel cell. Enzyme and Microbial Technology, 370 (2):0 238–245, 2005. ISSN 0141–0229. doi:10. 1016/j.enzmictec.2005.03.006

I.A. Ieropoulos, C. Melhuish, J. Greenman, I. Horsfield, Ecobot.ii: An artificial agent with a natural metaboloism. Int. J. Adv. Rob. Syst. **2**(4), 295–300 (2005b)

A.J. Ijspeert, A. Martinoli, A. Billard, L.M. Gambardella, Collaboration through the exploitation of local interactions in autonomous collective robotics: the stick pulling experiment. Auton. Robots **11**(2), 149–171 (2001)

A. Ilachinski, *Cellular Automata: A Discrete Universe* (World Scientific Publishing Company Incorporated, Singapore, 2001)

L. Iocchi, D. Nardi, M. Salerno, Balancing Reactivity and Social Deliberation in Multi-agent systems, chapter Reactivity and Deliberation: A Survey on Multi-robot Systems. Lecture Notes in Computer Science, vol. 2103 (Springer 2001), pp.9–32doi:10.1007/3-540-44568-4-2

R. Isermann, Process fault detection based on modeling and estimation methods?a survey. Automatica **20**(4), 387–404 (1984)

R. Isermann, Model-based fault-detection and diagnosis-status and applications. Annu. Rev. Control **29**(1), 71–85 (2005)

R. Isermann, P. Ballè, Trends in the application of model-based fault detection and diagnosis of technical processes. Control Engineering Practice, 50 5(0), pp. 709–719, 1997. ISSN 0967–0661. doi:10.1016/S0967-0661(97)00053-1

B.P. Jeppesen, D. Cebon, Analytical redundancy techniques for fault detection in an active heavy vehicle suspension. Veh. Sys. Dyn. **42**(1–2), 75–88 (2004)

B.W. Johnson, Fault-tolerant microprocessor-based systems. IEEE Micro **4**(6), 6–21 (1984)

L.K. Johnson, Foraging strategies and the structure of stingless bee communities in costa rica. Soc. Insects Tropics **1**, 31–58 (1982)

M.W. Jørgensen, E.H. Østergaard, H.H. Lund, Modular atron: Modules for a self-reconfigurable robot, in *Proceedings of 2004 IEEE/RSJ International Conference on Intelligent Robots and Systems (IROS)*, pp. 2068–2073, Sendai, Japan, 2004

M.R. Kabuka, S. Harjadi, A. Younis, A fault-tolerant architecture for an automatic vision-guided vehicle. IEEE Trans. Syst. Man Cybern. Cybern. **20**, 381–393 (1990). doi:10.1109/21.52549

A. Kamimura, H. Kurokawa, E. Yoshida, S. Murata, K. Tomita, S. Kokaji, distributed adaptive locomotion by a modular robotic system, m-tran ii, in *Proceddings of 2004 IEEE/RSJ International Conference on Intelligent Robots and Systems*, pp. 2370–2377, Sep. 2004

S.T. Kazadi, E. Kondo, A. Cheng, *A Robust Centralized Linear Spatial Search Flock* (ACTA Press, In Robotics and Applications, 2004)

I. Kelly, O. Holland, C. Melhuish, Slugbot: A robotic predator in the natural world, in *Proceedings of the 5th International Symposium on Artificial Life and Robotics*, pp. 470–475, 2000

J. Kennedy, R.C. Eberhart, Particle swarm optimization, in *IEEE International Conference on Neural Networks*, 1995

S. Kernbach, E. Meister, F. Schlachter, K. Jebens, M. Szymanski, J. Liedke, D. Laneri, L. Winkler, T. Schmickl, R. Thenius, P. Corradi, L. Ricotti, Symbiotic robot organisms: Replicator and symbrion projects, in *Proceedings of the 8th Workshop on Performance Metrics for Intelligent Systems, PerMIS '08* (ACM, New York, 2008) pp. 62–69, ISBN 978-1-60558-293-1. doi:10.1145/1774674. 1774685

S. Kernbach, L. Paul, E. Meister, F. Schlachter, O. Kernbach, Towards self-adaptation of robot organisms with a high developmental plasticity, in *The First International Conference on Adaptive and Self-Adaptive Systems and Applications, IEEE Computation World*, Nov. 2009

S. Kornienko, O. Kornienko, P. Levi, Ir-based communication and perception in micro-robotic swarms, in *Proceedings of the IEEE/RSJ International Conference on Intelligent Robots and Systems (IROS'05)*, Edmonton, Canada, Los Alamitos, CA, 2005

P.J.A.M. Korst, H.H.W. Velthuis, The nature of trophallaxis in honeybees. Insectes Soc. **29**(2), 209–221 (1992)

C.R. Kube, H. Zhang, Collective robot intelligence, in *Proceedings of the Second International Conference on From Animals to Animats 2: Simulation of Adaptive Behavior* (MIT Press, Cambridge 1993) pp. 460–468

T.H. Labella, M. Dorigo, J.-L. Deneubourg, Efficiency and task allocation in prey retrieval, in *BioADIT*, vol. 3141, Lecture Notes in Computer Science (LNCS), ed. by A.J. Ijspeert, M. Murata, N. Wakamiya (Springer, Heidelberg, 2004), pp. 274–289

G.A. Landis, P.P. Jenkins, Dust on mars: Materials adherence experiment results from mars pathfinder, in *Photovoltaic Specialists Conference, 1997, Conference Record of the Twenty-Sixth IEEE*, pp. 865–869, Sept. 1997 doi:10.1109/PVSC.1997.654224

W.F. Larsen, *Fault tree analysis* (Technical report, DTIC Document, 1974)

K. Lerman, C. Jones, A. Galstyan, M.J. Matarić, Analysis of dynamic task allocation in multi-robot systems. Inter. J. Rob. Res. **25**(3), 225–241 (2006)

M.L. Leuschen, J.R. Cavallaro, I.D. Walker, Robotic fault detection using nonlinear analytical redundancy, in *Robotics and Automation, 2002. Proceedings. ICRA'02. IEEE International Conference on*, vol. 1, pp. 456–463. IEEE, 2002

C. Lin, C. Lee, Fault-tolerant reconfigurable architecture for robot kinematics and dynamics computations. IEEE Trans. Syst. Man Cybern. Cybern. **21**, 983–999 (1991). doi:10.1109/21.120051

D. Linden, T.B. Reddy, *Handbook of Batteries*, 3rd edn. (McGraw-Hill, New York, 2002)

J.R. Lorch, A.J. Smith, Software strategies for portable computer energy management. IEEE Pers. Commun. **5**, 60–73 (1998). doi:10.1109/98.683740

W. Lui. Design and Modeling of Adaptive Foraging in Swarm Robotic Systems. PhD thesis, Faculty of Environment and Technology, University of the West of England, Bristol, Aug. 2008

W. Lui, A. Winfield, J.S. Chen, L. Dou, Strategies for energy optimization in a swarm of foraging robots. eds. by E. Şahin, W.M. Spears, A.F.T. Winfiel, Swarm Robotics: Second International Workshop, SAB 2006, LNCS, vol. 4433, Springer, Heidelberg, Sep. 2006

H.H. Lund, R. Beck, L. Dalgaard, Atron hardware modules for self-reconfigurable robotics, in Sugisaka and Takaga, editors, *Proceedings of the 10th International Symposium on Artificial Life and Robotics (AROB'10)*, Oita, Japan, 2005

A. Lyder, R. Garcia, K. Stoy, Mechanical design of odin, an extendable heterogeneous deformable modular robot, in *IEEE/RSJ International Conference on Intelligent Robots and Systems, (IROS'08)*, pp. 883–888, Nice, France, september 2008. doi:10.1109/IROS.2008.4650888

M.M. Mark. Swarms, Phase transitions and collective intelligence, ed. by C.G. Langton, *Artificial life* III, vol. XVII, pp. 417–445, MA, Jan 1994. Santa Fe Institute Studies in the Sciences of the Complexity, Addison-Wesley

A. Martinoli, K. Easton. Modeling swarm robotic systems. Experimental Robotics VIII, pp. 297–306, 2003

A. Martinoli, K. Easton, W. Agassounon, Modeling swarm robotic systems: a case study in collaborative distributed manipulation. Inter. J. Robot. Res. **23**(4–5), 415–436 (2004)

M.J. Matarić. Minimizing complexity in controlling a mobile robot population, in *Proceedings of the 1992 IEEE International Conference on Robotics and Automation*, pp. 830–835, Nice, France, 1992

M.J. Matarić, Reinforcement learning in the multi-robot domain. Auton. Robots **4**(1), 73–83 (1997)

M.J. Matarić, M.J. Marjanovic, Synthesizing complex behavior by composing simple primitives, in *Proceedings of the Self Organization and Life, From Simple Rules to Global Complexity*, vol. 2, pp. 698–707, Brussels, Belgium, 1993. European Conference on Artificial Life

M.J. Matarić, M. Nilsson, K.T. Simsarian, Cooperative multi-robot box-pushing, in *Proceedings of the International Conference on Intelligent Robots and Systems (IROS'95)*, vol. 3, pp. 3556-, Washington, DC, USA, 1995. IEEE Computer Society. ISBN 0-8186-7108-4

A. Matinoli, A.J. Ijspeert, L.M. Gambardella, A probabilistic model for understanding and comparing collective aggregation mechanisms, in *Proceedings of the 5th European Conference on Advances in Artificial Life (ECAL-99), vol. 1674, LNAI*, ed. by D. Floreano, J.D. Nicoud, F. Mondada (Springer, Berlin, 1999), pp. 575–584

D. McFarland, Autonomy and self-sufficiency in robots, in *The Artificial Life Route To Artificial Intelligence: Building Embodied Situated Agents*, ed. by L. Steels, R. Brooks (Lawrence Erlbaum, USA, 1994), pp. 187–213

D. McFarland, T. Boesser, *Intelligent Behavior in Animals and Robots* (MIT Press, Cambridge, 1993)

D. McFarland and E. Spier. Basic cycles, utility and opportunism in self-sufficient robots. Robotics and Autonomous Systems, 200 (2–4): 179–190, 1997. ISSN 0921–8890. doi:10.1016/S0921-8890(96)00069-3 Practice and Future of Autonomous Agents

Y. Mei, Y.-H. Lu, Y.C. Hu , C.S.G. Lee. A case study of mobile robots energy consumption and conservation techniques, in *Proceedings of the 12th International Conference Advanced Robotics (ICAR-05)*, pp. 492–497, 2005

C. Melhuish and M. Kubo. Collective energy distribution: Maintaining the energy balance in distributed autonomous robots using trophallaxis, eds. by R. Alami, R. Chatila, H. Asama, *Distributed Autonomous Robotic Systems 6* (Springer Japan, 2007) pp. 275–284. ISBN 978-4-431-35873-2

C.R. Melhuish, *Strategies for Collective Minimalist Mobile Robots* (Wiley, New Jersey., 2001)

R. Milne, Strategies for diagnosis. IEEE Trans. Syst. Man Cybern. Cybern. **17**(3), 333–339 (1987)

R. Moeckel, C. Jaquier, K. Drapel, E. Dittrich, A. Upegui, and A. Ijspeert. Yamor and bluemove an autonomous modular robot with bluetooth interface for exploring adaptive locomotion, eds by M.O. Tokhi, G.S. Virk, M.A. Hossain, *Climbing and Walking Robots* (Springer, Berlin Heidelberg, 2006), pp. 685–692. ISBN 978-3-540-26413-2. doi:10.1007/3-540-26415-9_82

A. Munoz, F. Sempé, A. Drogoul, R.G. Leclerc. Sharing a charging station in collective robotics. Technical report, Tech. Rep. LIP6 2002/026, 8, rue du Capitaine Scott, 75015 Paris, 2002

S. Murata, E. Yoshida, A. Kamimura, H. Kurokawa, M-tran: self-reconfigurable modular robotic system. IEEE/ASME Trans. Mechatron. **7**, 431–441 (2002)

T. Ngo, H. Schiøler, Truly autonomous robots: hardware design for an energy trophallactic robot. Artificial Life and Robotics, 12:0 335–345, 2008. ISSN 1433–5298. doi:10.1007/s10015-007-0495-z

T.D. Ngo, Towards Sociable Robots—Beyond Energy Autonomy. PhD thesis, Department of Electronic Systems, Aalborg University, Denmark, 2008

E. Nitz, R.C. Arking, T. Balch, Communication of behavioral state, in *multi-agent retrieval tasks, in Proceedings of the 1993 IEEE International Conference on Robotics and Automation, vol. 3*, ed. by L.W. Robert, O. Conner (IEEE Computer Society Press, Atlanta, GE, May 1993), pp. 588–594

G.F. Oster, E.O. Wilson, *Caste and Ecology in the Social Insects* (Princeton University Press, New Jersey, 1979)

E.H. Østergaard, Distributed control of the ATRON self-reconfigurable robot. PhD thesis, Maersk McKinney Moller Institute for Production Technology, University of Southern Denmark, Nov. 2004

N.D. Owens, J. Timmis, A.J. Greensted, A.M. Tyrell, On immune inspired homeostasis for electronic systems, eds by L. Castro, F. Zuben, H. Knidel, *Artificial Immune Systems*, vol. 4628 of *Lecture Notes in Computer Science* (Springer, Berlin Heidelberg, 2007), pp. 216–227, ISBN 978-3-540-73921-0. doi:10.1007/978-3-540-73922-7_19

R. Pablo, R. Ismael, R. Fernando, Using river formation dynamics to design heuristic algorithms, in *Unconventional Computation, Lecture notes in computer science*, pp. 163–177, (Springer Berlin, 2007)

L.E. Parker, The effect of action recognition and robot awareness in cooperative robotic teams. IEEE/RSJ International Conference on Intelligent Robots and Systems, 1:0 212, 1995. http://doi.ieeecomputersociety.org/10.1109/IROS.1995.525799

L.E. Parker, *Reliability and fault tolerance in collective robot systems (Fundamentals and Challenges, page To appear* (Pan Stanford Publishing, Handbook on Collective Robotics, 2012)

J.K. Parrish, S.V. Viscido, D. Grúnbaum, Self-organized fish schools: an examination of emergent properties. Biol. Bull. **202**(3), 296–305 (2002)

R.J. Patton, P.M. Frank, R. Clark, *Fault Diagnosis in Dynamic Systems: Theory and Applications* (Prentice Hall, Hertfordshire, 1989)

M.F. Paul, Fault diagnosis in dynamic systems using analytical and knowledge-based redundancy: A survey and some new results. Automatica, 260 3(0), pp. 459–474, 1990. ISSN 0005–1098. doi:10.1016/0005-1098(90)90018-D

D.W. Payton, D. Keirsey, D.M. Kimble, J. Krozel, J.K. Rosenblatt, Do whatever works: A robust approach to fault-tolerant autonomous control. Appl. Intell. **2**(3), 225–250 (1992)

D.C. Queller, J.E. Strassmann, Kin selection and social insects. BioScience **48**(3), 165–175 (1998)

W.C. Reynolds, Flocks, herds and school: a distributed behavioral model. Comput. Graphics **21**(4), 25–34 (1987)

B. Salemi, M. Moll, W.-M. Shen, Superbot: A deployable, multi-functional and modular self-reconfigurable robotic system, in *IEEE/RSJ Intl. Conf. Intelligent Robots Systems*, pp. 3636–3641, October 2006

T. Schmickl, K. Crailsheim, Trophallaxis within a robotic swarm: bio-inspired communication among robots in a swarm. Autonomous Robots, 25:0 171–188, 2008. ISSN 0929–5593

W.M. Shen, P. Will, P.A. Galstyan, C.M. Chuong, Hormone-inspired self-organization and distributed control of robotic swarms. Auton. Robots **17**(1), 93–105 (2004)

D.P. Siewiorek, D. Johnson, A design methodology for high reliability systems: The intel 432. The Theory and Practice of Reliable System Design, pp. 621–36, 1981

O. Soysal, E. Sahin, Probabilistic aggregation strategies in swarm robotic systems, in *Swarm Intelligence Symposium, 2005. SIS 2005. Proceedings 2005 IEEE*, pp. 325–332. IEEE, 2005

D.J.T. Sumpter, M. Beekman, From nonlineraity to optimality: pheromone trail foraging by ants. Anim. Behav. **66**, 273–280 (2003). doi:10.1006/anbe.2003.2224

H.J.C. Toby, A.M. Bruce, Player 2.0: Toward a practical robot programming framework, in *Proceedings of the Australasian Conference on Robotics and Automation (ACRA)*, 2005

H. Torres-Contreras, R. Olivares-Donoso, H.M Niemeyer, Solitary foraging in the ancestral south american ant, pogonomyrmex vermiculatus. is it due to constraints in the production or perception of trail pheromones? Journal of chemical ecology (J. Chem. Ecol.), 330 2(0), pp. 435–40, 2007. ISSN 0098–0331. http://www.biomedsearch.com/nih/Solitary-foraging-in-ancestral-South/17187299.html

V. Trianni, M. Dorigo, Emergent collective decisions in a swarm of robots, in *Swarm Intelligence Symposium, 2005. SIS 2005. Proceedings 2005 IEEE*, pp. 241–248. IEEE, 2005

V. Trianni, T.H. Labella, R. Gross, E. Sahin, M. Dorigo, J.L. Deneubourg, Modeling pattern formation in a swarm of self-assembling robots. European community grant 1ST-2000-31010, 2002

V. Trianni, R. Groß, T. Labella, E. Şahin, M. Dorigo, Evolving aggregation behaviors in a swarm of robots. Adv. Artif. Life **78**, 865–874 (2003)

V. Trianni, S. Nolfi, M. Dorigo, Cooperative hole avoidance in a swarm-bot. Robot. Auton. Syst. **54**(2), 97–103 (2006)

P. Ulam, T. Balch, Using optimal foraging models to evaluate learned robotic foraging behavior. Adapt. Behav. Animals, Animats, Software Agents, Robots, Adaptive Syst. **12**(3–4), 213–222 (2004)

J.D. Villa, Swarming behavior of honey bees (hymenoptera: Apidae) in southeastern louisiana. Ann. Entomol. Soc. Am. **97**(1), 111–116 (2004)

U.B. Wade, C.M. Gifford, Investigation of power sources for the polar seismic tetwalker. Technical report, Centre for Remote Sensing of Ice Sheets (CReSIS), University of Kansas, Lawrence, 2007

A.J. Wainselboim, W.M. Farina, Trophallaxis in honeybees, apis mellifera (l.), as related to their past experience at the food source. Animal Behaviour, 660 (4):0 791–795, 2003. ISSN 0003–3472. doi:10.1006/anbe.2003.2256

W.M. Wheeler, he ant-colony as an organism. J. Morphol. **22**(2), 307–325 (1911)

W.M. Wheeler, *Social Life Among the Insects* (antbase.org, Harcourt, Brace and Co., New York, 1923)

E.P. Widmaier, R. Hershel, K.T. Strang, Vander's Human Physiology: The Mechanisms of Body Function. McGraw-Hill, Ney York, 10th edition, 2006. ISBN 0-07-111677-X

S. Wilkinson, Gastrobots: Benefits and challenges of microbial fuel cells in food powered robot applications. Autonomous Robots, 9:0 99–111, 2000. ISSN 0929–5593

A.S. Willsky, A survey of design methods for failure detection in dynamic systems. Automatica **12**, 601–611 (1976). doi:10.1016/0005-1098(76)90041-8

E. Wilson, T. Eisner, Quantitative studies of liquid food transmission in ants. Insectes Sociaux, 4(0), pp. 157–166, 1957. ISSN 0020–1812. doi:10.1007/BF02224149

E.O. Wilson, B. Hölldobler, Dense heterarchies and mass communication as the basis of organization in ant colonies. Trends Ecol. Evol. **3**, 65–68 (1988)

A. Winfield, J. Nembrini, Safety in numbers: fault-tolerance in robot swarms. Int. J. Model. Ident. Control **1**(1), 30–37 (2006)

L. Winkler, H. Wörn, Symbricator3d - a distributed simulation environment for modular robots, eds. by M. Xie, Y. Xiong, C. Xiong, H. Liu, Z. Hu, Intelligent Robotics and Applications, vol. 5928 of Lecture Notes in Computer Science (Springer, Berlin, 2009) pp. 1266–1277. ISBN 978-3-642-10816-7

L. Winkler, Vonasek V., W. Heinz, P. Libor, Robot3d—a simulator for mobile modular self-reconfigurable robots, in *IEEE Conference on Multisensor Fusion and Integration for Intelligent Systems (MFI)*, 2012, pp. 464–469

P. Xuan, V. Lesser, S. Zilberstein, Communication decisions in multi-agent cooperation: model and experiments, in *Proceedings of the fifth international conference on Autonomous agents, AGENTS '01*(ACM, New York, 2001) pp. 616–623, ISBN 1-58113-326-X. doi:10.1145/375735.376469

M. Yim, D.G. Duff, K.D. Roufas' Polybot: a modular reconfigurable robot, in *Proceedings of the 2000 IEEE International Conference on Robotics & Automation*, pp. 514–520. Xerox Palo Alto research Center, Apr. 2000

V. Zykov, A. Chan, H. Lipson, Molecubes: An open-source modular robotics kit, in *Proceedings of the International Conference on Intelligent Robots and Systems (IROS-07), IEEE/RSJ Workshop on Self-Reconfigurable Robotics.,* 2007

MIX
Papier aus verantwortungsvollen Quellen
Paper from responsible sources
FSC® C105338
FSC
www.fsc.org

If you have any concerns about our products,
you can contact us on
ProductSafety@springernature.com

In case Publisher is established outside the EU,
the EU authorized representative is:
Springer Nature Customer Service Center GmbH
Europaplatz 3, 69115 Heidelberg, Germany

Printed by Libri Plureos GmbH
in Hamburg, Germany